Graphene Optics:
Electromagnetic Solution of
Canonical Problems

Graphene Optics: Electromagnetic Solution of Canonical Problems

Ricardo A Depine
Department of Physics, University of Buenos Aires, Argentina

Morgan & Claypool Publishers

Rights & Permissions
To obtain permission to re-use copyrighted material from Morgan & Claypool Publishers, please contact info@morganclaypool.com.

ISBN   978-1-6817-4309-7 (ebook)
ISBN   978-1-6817-4308-0 (print)
ISBN   978-1-6817-4311-0 (mobi)

DOI   10.1088/978-1-6817-4309-7   .

Version: 20161201

IOP Concise Physics
ISSN 2053-2571 (online)
ISSN 2054-7307 (print)

A Morgan & Claypool publication as part of IOP Concise Physics
Published by Morgan & Claypool Publishers, 40 Oak Drive, San Rafael, CA, 94903 USA

IOP Publishing, Temple Circus, Temple Way, Bristol BS1 6HG, UK

*To all the members of the Applied Electromagnetic Group (UBA)
that I had the pleasure of working with, and from whom I have always
learned more than I have been able to teach.*

Contents

Preface

Silicon, the chemical element with atomic number 14, has become the most popular material in the information-technology revolution. Gordon Moore's observation that transistors were shrinking so fast that their number on a microprocessor chip per unit area would double over a period of time $\approx two$ years (Moore's law) governed the exponential improvement that made it possible to have smartphones that are millions of times more powerful than all the combined computing that in 1969 allowed humans to travel through space, land on the Moon and return safely to the Earth.

Today it is clear that the doubling of Moore's law is coming to an end. With faster and faster electrons moving through smaller and smaller silicon circuits, overheating has been found to be a problem, and when chips began to get too hot the beginning of the end started. The problem has been temporarily solved by continuing with the doubling, but limiting the speed of the processors in order not to generate too much heat. However, if the doubling continues, in less than a decade the size of transistors will reach the limit of 2–3 nanometers, that is, the size of ten atoms in line. In this scale, quantum uncertainties will affect electron behavior, thus making transistors unreliable.

The search for alternatives to today's silicon technology has driven research in areas such as quantum computing, spintronics and low-dimensional materials. This book deals with the third area of research mentioned above, and is devoted to the interaction of electromagnetic radiation with graphene, a single atomic sheet of carbon in a hexagonal lattice, considered the first strictly two-dimensional material. The isolation and characterization of graphene in 2004, which resulted in the Nobel prize in physics being awarded to Andre Geim and Konstantin Novoselov in 2010, fueled the discovery of other two-dimensional materials, such as boron nitride, transition metal and gallium dichalcogenides, and led to an explosion of research and development activity in the electronics and photonics of flat materials and devices with the potential to provide alternatives to silicon technology.

Apart from its remarkable thinness, graphene is a zero-band-gap semiconductor with conductivity tuned by either electrostatic or magnetostatic gating, and it can support highly confined surface plasmons with tunable dispersion. The remarkable electronic and photonic properties of graphene make it attractive for the development of tunable nanoelectronic and photonic devices that are potentially relevant for on-chip elements such as light sources, photodetectors, sensors or optical modulators. While many graphene applications have been designed to exploit the planar ultrathin configuration, others rely on the properties that emerge when graphene sheets are stacked into structures that are still very thin, but definitely 3D.

This book was written in the same way I would teach a course at graduate or advanced undergraduate level on rigorous macroscopic description of the interaction between electromagnetic radiation and structures containing graphene sheets, and assuming a course in electrodynamics developed from Maxwell's equations as a prerequisite. To comply with the purpose of the IOP Concise

Physics collection—keeping the text short while providing an introduction to the topic—I have chosen to present only canonical problems with translational invariant geometries, in which the solution of the original vectorial problem can be reduced to the treatment of two scalar problems, corresponding to two basic polarization modes. This choice excludes other significant problems, such as the Mie-like solution for the scattering of electromagnetic radiation by a graphene sphere, which can be considered the paradigm of a 3D graphene particle, but includes the analogous solution for the scattering of electromagnetic radiation by a graphene wire of circular section, which can be considered the paradigm of a 2D graphene particle.

The first chapter provides a summary of the topics of macroscopic electro-magnetism which are essential to understand the approach used to model the electromagnetic response of graphene, such as the distinction between free versus bound charges and currents, the formulation of constitutive relations in the frequency domain, the constitutive properties of graphene described by the Kubo model for the graphene surface conductivity and the boundary conditions in the presence of graphene layers. The electromagnetic analysis of graphene structures associated with different degrees of symmetry is given in subsequent chapters, starting from geometries exhibiting two-dimensional continuous translational symmetry, such as the single flat and homogeneously doped sheet (the Fresnel problem, chapter 2) and the graphene planar waveguide (chapter 3), and then moving on to geometries with two-dimensional continuous translational symmetry along the graphene sheets and one-dimensional discrete translational symmetry along the normal direction (1D photonic crystals, section 3.4), before continuing with geometries with one-dimensional continuous translational symmetry along one tangent direction on the graphene sheet and one-dimensional discrete translational symmetry along the other (perpendicular) tangent direction on the graphene sheet (graphene gratings, chapter 4), before finally arriving at structures that keep the one-dimensional continuous translational symmetry along the wire axis, but do not have any translational symmetry in the plane perpendicular to this axis (graphene covered wires, chapter 5). Taking into account the fact that since the advent of photonic metamaterials exhibiting magnetism at high frequencies, many well-established optic laws for conventional (nonmagnetic) materials now have an exception, the theoretical expressions in this book are given for materials with non-unity magnetic permeability. However, for reasons of concision, very interest-ing phenomena, such as negative refraction and propagation in zero index or backward surface waves, have not been investigated in the examples.

I really enjoy computational physics and I always try to include computational problems in my courses. In order to study the nature of the Universe, a physicist needs numerical answers and although in the past it was not easy for non-programmers to get into some type of computing environment, nowadays we have excellent tools, such as the Python language, which makes numerical calculations accessible to any science student, even those who have not attended a computer science class. Therefore, many figures in the book are accompanied by Python scripts, which I hope will generate discussions about physics, rather than programming.

I hope the book will be useful for advanced undergraduate students, postgraduate students and postdocs interested in the rigorous macroscopic description of the interaction between electromagnetic radiation and two-dimensional materials. Comments and suggestions will be very much appreciated.

Acknowledgments

I wish to thank Mauro Cuevas for encouraging me to explore the graphene world, Máximo Riso for his efforts in revising lengthy algebraic procedures and my wife Begoña for always being willing to help, with the English language and all kinds of everyday life matters.

Author biography

Ricardo A Depine

Ricardo A Depine was born in Buenos Aires, Argentina. He is a Permanent Professor with the Department of Physics at the University of Buenos Aires (UBA), where he leads the research activities of the Applied Electromagnetics Group and teaches courses on classical electrodynamics, waves, photonics and optics and continuum mechanics. He is also Principal Researcher of the Consejo Nacional de Investigaciones Científicas y Técnicas (CONICET). His current research activities encompass the electromagnetic response of nanostructures, scattering by rough surfaces, photonic crystals, materials and photonics, plasmonics, and metamaterials. More details about his current and past research interests can be found at http://users.df.uba.ar/rdep/.

Graphene Optics: Electromagnetic Solution of
Canonical Problems

Ricardo A Depine

Chapter 1

Electromagnetics of graphene

This chapter summarizes the aspects of macroscopic electromagnetism that form the basis of this book. The emphasis is placed on topics that are essential to understand the approach used to model the electromagnetic response of graphene, such as the distinction between the bound charges and currents associated with atoms or molecules and the free charges and currents associated with external sources; how the bound sources are described in terms of polarization and magnetization densities; the formulation of constitutive relations in the frequency domain; the Kubo model for graphene surface conductivity, the form of the fields in piecewise homogeneous materials; and the boundary conditions in the presence of a graphene layer.

1.1 Macroscopic electrodynamics

The electromagnetic field can be described as the combination of electric and magnetic fields produced by charges and currents, i.e. electrically charged moving objects. Due to the microscopic nature of matter, all matter is an ensemble of discrete charges dispersed in free space or vacuum, and bound charges and currents are induced in a material in the presence of electromagnetic fields. Macroscopic electrodynamics can be formulated starting from the fundamental laws of electromagnetism in vacuum. However, this formulation involves the description of the very complicated and granular microscopic sources, and the drastic variations in space they produce on the microscopic fields. In order to avoid such problems, macroscopic electrodynamics is instead formulated in terms of averaged fields and sources, which describe the behavior of fields and sources measured with an instrument of macroscopic dimensions. All averages are performed over a sufficiently large scale, much larger than individual atomic sizes so as not to see the granularity of the atomic sources, but also sufficiently small compared to the distances accessible to experimentation [1].

We start with macroscopic Maxwell's equations in SI units:

$$\nabla \times \boldsymbol{E}(\boldsymbol{r}, t) = -\frac{\partial \boldsymbol{B}}{\partial t}(\boldsymbol{r}, t) \tag{1.1}$$

$$\nabla \times \boldsymbol{H}(\boldsymbol{r}, t) = \frac{\partial \boldsymbol{D}}{\partial t}(\boldsymbol{r}, t) + \boldsymbol{J}(\boldsymbol{r}, t) \tag{1.2}$$

$$\nabla \cdot \boldsymbol{D}(\boldsymbol{r}, t) = \rho(\boldsymbol{r}, t) \tag{1.3}$$

$$\nabla \cdot \boldsymbol{B}(\boldsymbol{r}, t) = 0, \tag{1.4}$$

where $\boldsymbol{r} = (x, y, z)$, t denotes the time, $\boldsymbol{E}$ denotes the electric field, $\boldsymbol{D}$ denotes the electric displacement, $\boldsymbol{B}$ and $\boldsymbol{H}$ denote the magnetic fields, and ρ and $\boldsymbol{J}$ denote the free charge and current volume densities, respectively, that is, the free sources not associated with any particular atom or molecule. Since magnetic terminology can be confusing—magnetic induction and magnetic field are used in the literature to describe either $\boldsymbol{B}$ or $\boldsymbol{H}$ without distinction—conventional names for $\boldsymbol{B}$ and $\boldsymbol{H}$ are avoided in this book. Taking into account the fact that $\boldsymbol{E}$ and $\boldsymbol{B}$ appear in the Lorentz force acting on a point charge $q(\boldsymbol{r}, t)$ travelling at velocity $\boldsymbol{v}(\boldsymbol{r}, t)$,

$$\boldsymbol{F}_{\text{Lor}} = q(\boldsymbol{r}, t)[\boldsymbol{E}(\boldsymbol{r}, t) + \boldsymbol{v}(\boldsymbol{r}, t) \times \boldsymbol{B}(\boldsymbol{r}, t)], \tag{1.5}$$

the fundamental electromagnetic quantity, $\boldsymbol{E}$ and $\boldsymbol{B}$, may be therefore called primitive fields, while $\boldsymbol{D}$ and $\boldsymbol{H}$ are macroscopic fields induced by the presence of material interactions and may be appropriately termed induction fields.

Macroscopically averaged bound charges and currents are described in terms of the polarization density $\boldsymbol{P}(\boldsymbol{r}, t)$, representing the electric dipole moment per unit volume, and the magnetization density $\boldsymbol{M}(\boldsymbol{r}, t)$, representing the magnetic dipole moment per unit volume. The macroscopic bound charge density $\rho_{\text{b}}(\boldsymbol{r}, t)$ and bound current density $\boldsymbol{J}_{\text{b}}(\boldsymbol{r}, t)$ are

$$\rho_{\text{b}} = -\nabla \cdot \boldsymbol{P} \tag{1.6}$$

$$\boldsymbol{J}_{\text{b}} = \nabla \times \boldsymbol{M} + \frac{\partial \boldsymbol{P}}{\partial t}. \tag{1.7}$$

1.2 Constitutive relations

The use of $\boldsymbol{D}$ and $\boldsymbol{H}$ has the advantage that only the free sources, not the bound ones, appear explicitly in equations (1.2) and (1.3), that is, in Maxwell equations involving the sources. However, the information about the bound sources given by equations (1.6) and (1.7) is of course not irrelevant. In fact, P and M appear in the definition of D and H:

$$D = \varepsilon_0 E + P \tag{1.8}$$

$$H = \mu_0^{-1} B - M, \tag{1.9}$$

where ε_0 and μ_0 are, respectively, the electric permittivity and magnetic permeability of vacuum, $\sqrt{\varepsilon_0 \mu_0} = c^{-1}$, and c is the velocity of light in vacuum. The behavior of P and M under the influence of the fields results from the microscopic structure of the material. The equations describing this behavior are known as constitutive relations. In some cases the determination of P and M in terms of E and B can be derived from first principles, using the tools of condensed matter physics, whereas in other cases this determination is based directly upon measurements. Constitutive relations can be symbolically expressed in the following form [1]:

$$D = D[E, B] \tag{1.10}$$

$$H = H[E, B]. \tag{1.11}$$

For conducting media, i.e. media that generate an electric current in the presence of an electric field, it is sometimes convenient to split J, the current volume density not associated with any particular atom or molecule, into an external current density $J^{(\text{ext})}$ that is associated with external sources, and an induced current density $J^{(\text{cond})}$ that is associated with the conduction carriers:

$$J = J^{(\text{cond})} + J^{(\text{ext})}. \tag{1.12}$$

In this case, a generalized Ohm's law

$$J^{(\text{cond})} = J^{(\text{cond})}[E, B] \tag{1.13}$$

must also be given. Square brackets were used to denote that the connections between the induced and the primary quantities are not necessarily simple. For example, these connections may include the dependence of D and H (or equivalently, of P and M) on E and B at other locations and times. The study of constitutive relations provides a basis for the classification of media using certain criteria, such as distinctions on the basis of linearity versus nonlinearity, isotropy versus anisotropy, homogeneity versus nonhomogeneity, spatial and/or temporal dispersion, and so on [2]. Some of the most common materials have a linear response to applied fields over a wide range of field values. The constitutive relations for a *linear* medium are generally nonlocal with respect to both space and time and can be written as [2, 3]:

$$\begin{aligned} D(r, t) = \int d^3 r' \, dt' \Big[\check{\varepsilon}_{\text{EB}}(r', t') \cdot E(r - r', t - t') \\ + \check{\xi}_{\text{EB}}(r', t') \cdot B(r - r', t - t') \Big] \end{aligned} \tag{1.14}$$

$$H(r, t) = \int d^3r' \, dt' \Big[\check{\xi}_{EB}(r', t') \cdot E(r - r', t - t')$$
$$+ \check{\nu}_{EB}(r', t') \cdot B(r - r', t - t') \Big], \tag{1.15}$$

where $\check{\varepsilon}_{EB}(r, t)$, $\check{\xi}_{EB}(r, t)$, $\check{\zeta}_{EB}(r, t)$ and $\check{\nu}_{EB}(r, t)$ are second order tensors. Temporal nonlocality refers to the fact that the induction fields D and H (and therefore the induced sources) at time t depend on the values of E and B at all times t' previous to t. Similarly, spatial nonlocality refers to the fact that D and H at location r depend on the values of E and B at neighboring locations r'. While temporal nonlocality, a consequence of causality [4], is a widely encountered phenomenon that must be accurately taken into account, spatial nonlocality can play a significant role when some characteristic length scale in the medium is short with respect to the wavelength, but can usually be neglected for visible light or electromagnetic radiation of longer wavelengths [5]. Restricting ourselves to linear, spatially local materials, the constitutive relations take the form:

$$D(r, t) = \int dt' \Big[\check{\varepsilon}_{EB}(r, t') \cdot E(r, t - t') + \check{\xi}_{EB}(r, t') \cdot B(r, t - t') \Big] \tag{1.16}$$

$$H(r, t) = \int dt' \Big[\check{\zeta}_{EB}(r, t') \cdot E(r, t - t') + \check{\nu}_{EB}(r, t') \cdot B(r, t - t') \Big]. \tag{1.17}$$

1.3 From the time domain to the frequency domain

In equations (1.14) and (1.15), spatial and temporal nonlocality are displayed as convolutions between constitutive tensors and fields. Thanks to the convolution theorem [6], constitutive equations are more conveniently handled in the domain of the angular frequency ω and the wavevector k, the Fourier conjugate variables for the time coordinate t and the space coordinate r, respectively. For spatially local materials, only the angular frequency ω is relevant. After taking the temporal Fourier transforms of equations (1.16) and (1.17), the following frequency-domain constitutive relations can be obtained

$$D_\omega(r) = \varepsilon_{EB}(r, \omega) \cdot E_\omega(r) + \xi_{EB}(r, \omega) \cdot B_\omega(r) \tag{1.18}$$

$$H_\omega(r) = \zeta_{EB}(r, \omega) \cdot E_\omega(r) + \mu_{EB}(r, \omega) \cdot B_\omega(r), \tag{1.19}$$

with $E_\omega(r)$, $B_\omega(r)$, $D_\omega(r)$ and $H_\omega(r)$ being the temporal Fourier transforms of the electromagnetic fields, and $\varepsilon_{EB}(r, \omega)$, $\mu_{EB}(r, \omega)$, $\xi_{EB}(r, \omega)$ and $\zeta_{EB}(r, \omega)$ being the temporal Fourier transforms of the constitutive tensors. Here we follow the convention that the inverse temporal Fourier transform incorporates a $-i$ factor in the exponential term, so a time-dependent quantity $F(r, t)$ in terms of its Fourier transform is obtained as

$$F(r, t) = \int_{-\infty}^{\infty} F_\omega(r) e^{-i\omega t} d\omega. \tag{1.20}$$

The form of equations (1.18) and (1.19), known as the Boys–Post representation [3], denotes the natural roles of E and B as the fundamental or primitive fields and D and H as the derived or induction fields. However, in common practice the Tellegen representation [3]

$$D_\omega(r) = \varepsilon_{EH}(r, \omega) \cdot E_\omega(r) + \xi_{EH}(r, \omega) \cdot H_\omega(r) \tag{1.21}$$

$$B_\omega(r) = \zeta_{EH}(r, \omega) \cdot E_\omega(r) + \mu_{EH}(r, \omega) \cdot H_\omega(r) \tag{1.22}$$

is usually preferred, with $\varepsilon_{EH}(r, \omega)$, $\mu_{EH}(r, \omega)$, $\xi_{EH}(r, \omega)$ and $\zeta_{EH}(r, \omega)$ being the temporal Fourier transforms of the constitutive tensors $\breve{\varepsilon}_{EH}(r, t)$, $\breve{\xi}_{EH}(r, t)$, $\breve{\zeta}_{EH}(r, t)$ and $\breve{\nu}_{EH}(r, t)$.

1.4 Isotropic media

For most materials P and $J^{(cond)}$ are parallel to E and M is parallel to B. These materials are known as isotropic dielectric–magnetic materials and their constitutive relations have the simple form

$$D_\omega(r) = \varepsilon_0 \, \varepsilon(r, \omega) E_\omega(r) \tag{1.23}$$

$$B_\omega(r) = \mu_0 \, \mu(r, \omega) H_\omega(r) \tag{1.24}$$

$$J_\omega^{(cond)}(r) = \sigma^{3D}(r, \omega) \, E_\omega(r), \tag{1.25}$$

where σ^{3D} is the electric conductivity, ε is the scalar electric permittivity and μ is the scalar magnetic permeability. For passive media, and as a consequence of the sign convention used for the exponential term in equation (1.20), the complex valued constitutive scalars ε and μ lie on the upper half of the complex plane, whereas σ^{3D} lies on the right half of the complex plane.

The classical vacuum is both isotropic and homogeneous. It is the reference medium of classical electromagnetism and its time-domain constitutive relations emerge from the definition of D and H when all the material interactions are removed. In this case, $P = 0$, $M = 0$ and $J^{(cond)} = 0$, and from equations (1.8) and (1.9) we obtain

$$D(r, t) = \varepsilon_0 \, E(r, t) \tag{1.26}$$

$$B(r, t) = \mu_0 \, H(r, t). \tag{1.27}$$

As ε_0 and μ_0 are true constants, the corresponding frequency-domain constitutive relations have the same form as equations (1.23) and (1.24), but with $\varepsilon(r, \omega) = 1$, $\mu(r, \omega) = 1$ and $\sigma^{3D}(r, \omega) = 0$.

1.5 Conductivity and permittivity formulations

We observe that the right-hand side in Ampère's law equation (1.2) has contributions associated with current densities of different natures. As shown from the definition of D given by equation (1.8), the first term $\partial D/\partial t$ gives one contribution that is proportional to the rate of change of the electric field E, which is present even in vacuum, and another contribution that is proportional to the rate of change of the electric polarization P, which represents polarization currents associated with oscillating bound charges and is given by the second term in equation (1.7). According to equation (1.12), the second term on the right-hand side in Ampère's law equation (1.2), representing the free current density, can also be separated into two other contributions, one from the external sources and the other from induced carriers.

To work with the easy form of the constitutive relations, it is now convenient to consider the frequency-domain version of equation (1.2). Doing so and using the constitutive equation (1.23) for D_ω, the separation equation (1.12), and the constitutive equation (1.25) for $J_\omega^{(cond)}$, we obtain

$$\nabla \times H_\omega(r) = -i\omega\varepsilon_0\left[\varepsilon(r,\, \omega) + \frac{i\sigma^{3D}(r,\, \omega)}{\omega\varepsilon_0}\right]E_\omega(r) + J_\omega^{(ext)}(r). \qquad (1.28)$$

The right-hand side of this equation shows that all the contributions to current density that are not associated with an external source are represented, for linear media, by a term that is proportional to the temporal Fourier transform of the electric field $E_\omega(r)$. In addition, it shows that the contribution from the displacement currents coming from $\partial D/\partial t$ and that from the conduction carriers coming from Ohm's law are both treated on the same footing through the factor in square brackets in equation (1.28)

$$\varepsilon^{(eff)}(r,\, \omega) = \left[\varepsilon(r,\, \omega) + \frac{i\sigma^{3D}(r,\, \omega)}{\omega\varepsilon_0}\right], \qquad (1.29)$$

which represents an effective dielectric constant to which all the constitutive properties of the medium can be attributed. We observe that this effective dielectric constant does not distinguish between conduction currents and polarization currents, which reflects the physical fact that no fundamental difference between the conductors and dielectrics exists for oscillatory fields. This is the case because, as the resulting oscillatory motion of all charges is spatially localized, the separation of the current into a polarization part and a free part is not really possible and the only combination that can ever matter is the sum [7].

Both the constitutive parameter associated with bound polarization charges and the effective dielectric constant represented by equation (1.29) are usually denoted by the same symbol, $\varepsilon(r,\, \omega)$. This ambiguous notation should not lead us to forget that, when using an effective dielectric constant or a dielectric constant and a conductivity, we are using two different formulations of electromagnetism which differ regarding how the conduction carriers are treated. The fundamental or primitive fields E_ω and B_ω in these two formulations are the same, but the derived or auxiliary fields D_ω and H_ω are different [7, 8].

1.6 Constitutive relations of graphene

Graphene is a single layer of a graphite crystal. It consists of a single atomic layer of pure covalently bonded carbon atoms arranged in a two-dimensional (2D) hexagonal lattice structure. Each carbon atom has six electrons surrounding its nucleus, two in the inner shell and four in the outer electron shell. Of these four electrons, three are bound with the nearest-neighbor atom electrons and create the strong chemical bonds that make graphene one of the strongest materials known to man, whereas the other electron in the outer electron shell of each carbon atom is delocalized on the whole graphene layer [9].

The energy band structure of the delocalized electron determines graphene's conductivity. While in non-conducting or semiconducting materials the full valence band and the empty conduction band are separated by an energy gap, there is no gap between the conduction and valence bands in pure graphene. Thus, pure graphene can be regarded as a zero-gap semiconductor. Similarly, while in metals the valence band is partially filled, in pure graphene the Fermi level lies at the point where the conduction and valence bands meet (the Dirac point). Thus, pure graphene can be regarded as a metal with an empty valence band. This duality between the zero-gap-semiconductor picture and the metal-with-an-empty-valence-band picture makes graphene particularly interesting for applications in many photonic devices that require conducting but transparent thin films.

The duality between semiconductor and metallic behavior can be controlled by tuning the position of the Fermi level from the Dirac point, either by chemical additions, or, more easily, by using the electric field effect [10]. For example, when a constant voltage (the *gate voltage*) is applied between graphene and a metallic layer, separated by a very thin insulator, the resulting electrical field will modify the quantity of conduction carriers and thus graphene's electrical conductivity. For a given polarity of the voltage, that is, for a given direction of the dc electric field, the conduction band fills up, which means that electrons are added to the system. For the other polarity, the number of electrons in the valence band is reduced, which means that holes are added to the system.

Graphene constitutive equations can be well described using the Schrödinger-equation-based methods of condensed matter physics. In 1947, more than half a century before its deliberate fabrication, the electronic structure of graphene was explored and its unusual properties were demonstrated by P R Wallace [11]. The motivation of his work was to use a single-layer formalism as a starting point to determine the electronic properties of bulk graphite. Using a band theory of solids with a tight-binding approximation, Wallace was able to explain many of the physical properties of graphite.

A relevant result from Wallace's pioneering study is the fact that in graphite electronic currents only take place in layers, in other words, along a graphene layer.

Therefore, a logical choice for incorporating a graphene layer into Maxwell's equations is to model it using concepts such as conductivity surface, sheet resistance or surface impedance. These concepts are widely applied in microwave electronics and antenna theory when thin films are considered as 2D entities [12, 13], or when the current $J^{(\mathrm{cond})}$ in conducting boundaries is confined to such a small thickness just below the surface of the conductor that is equivalent to an effective surface current K ([1], pp 354–6). For a linear and isotropic conductivity surface, the coefficient of proportionality $\sigma(\omega)$ linking the surface current and the tangential component of the electric field along the plane of the sheet is called *surface conductivity*. In this case, the 2D analogue of the constitutive equation (1.25) is written as

$$K_\omega = \sigma(\omega) E_\omega. \tag{1.30}$$

The inverse of $\sigma(\omega)$, the coefficient of proportionality linking the tangential component of the electric field and the surface current along the plane of the sheet, is usually called *surface impedance* ([1], pp 354–6, [14]). In this book we deal with situations in which graphene exhibits a local, linear and isotropic response.

As an alternative to the infinitesimally thin current sheet model, a homogeneous film of finite thickness δ can also be used to simulate a graphene layer [15], with the current K_ω evenly spread through the thickness of the film. The volume current density $J_\omega^{(\mathrm{cond})}$ is then $J_\omega^{(\mathrm{cond})} = K_\omega/\delta$, and from equation (1.25) it follows that

$$\sigma^{3\mathrm{D}}(\omega) = \sigma(\omega)/\delta. \tag{1.31}$$

Therefore, according to equation (1.29), the effective dielectric constant of this homogeneous film is

$$\varepsilon(\omega) = \varepsilon_0 \left[1 + \frac{i\sigma(\omega)}{\omega\varepsilon_0\delta} \right]. \tag{1.32}$$

1.7 Ohmic losses

Recalling the constitutive equation (1.30) and taking into account the fact that the time-averaged rate of work done per unit time per unit volume by the electromagnetic fields on a current volume distribution is $\mathrm{Re}\,\{J_\omega \cdot E_\omega^*\}/2$ [1], it follows that the time-averaged rate of work done per unit time per unit *surface* by the electromagnetic fields on the graphene charge carriers is

$$\frac{\mathrm{d}P}{\mathrm{d}a} = \frac{1}{2}\mathrm{Re}\left\{K_\omega\,E_\omega^*\right\} = \frac{1}{2}\mathrm{Re}\{\sigma\,|E_\omega|^2\} = \frac{\mathrm{Re}\,\sigma}{2}\,|E_\omega|^2, \tag{1.33}$$

where the asterisk is used to denote the complex conjugate of a complex number and we have used the notation $\mathrm{d}P/\mathrm{d}a$ to recall that this quantity is the time-averaged power transferred from the electromagnetic field to the graphene carriers per unit area.

1.8 Kubo model for the surface conductivity

Both the conductivity surface and the thin film model for a graphene sheet require knowledge of the frequency-dependent surface conductivity σ, a quantity which can be obtained from either a microscopic model [13, 16–25] or from measurements [26–28]. Here we use the following high-frequency expression derived from the Kubo model [13, 22, 25]

$$\sigma(\omega) = \frac{e^2(\omega + i\gamma_c)}{i\pi\hbar^2}\left[\int_{-\infty}^{+\infty} \frac{|\epsilon|}{(\omega + i\gamma_c)^2}\frac{df_0(\epsilon)}{d\epsilon}d\epsilon\right.$$
$$\left. - \int_0^{+\infty} \frac{f_0(-\epsilon) - f_0(\epsilon)}{(\omega + i\gamma_c)^2 - 4(\epsilon/\hbar)^2}d\epsilon\right], \tag{1.34}$$

where $-e$ is the electron charge, $\hbar = h/2\pi$ is the reduced Planck constant, γ_c is a phenomenological carrier scattering rate that is assumed to be independent of the energy ϵ, $f_0(\epsilon) = \{\exp[(\epsilon - \mu_c)/k_BT] + 1\}^{-1}$ is the Fermi function, μ_c is the chemical potential (controlled with the help of a gate voltage), k_B is Boltzmann's constant and T is the ambient temperature. The first term in equation (1.34) corresponds to intraband electron–photon scattering processes, whereas the second term corresponds to the direct interband electron transitions.

For undoped (no chemical additions) and ungated (zero gate voltage) graphene at $T = 0$ K, the charge carrier density n_0 is very low, but it can be tuned by chemical additions (doping) or with the help of a constant electric field (electric field effect, gate voltage). The chemical potential is determined through the condition

$$n_0 = \frac{2}{\pi\hbar^2 v_F^2}\int_0^\infty \epsilon\big[f_0(\epsilon) - f_0(\epsilon + 2\mu_c)\big]d\epsilon, \tag{1.35}$$

where $v_F \approx 9.5 \cdot 10^5$ m s^{-1} is the Fermi velocity.

1.8.1 Intraband conductivity

The intraband term in equation (1.34) results in

$$\sigma^{intra}(\omega) = \frac{2ie^2 k_B T}{\pi\hbar^2(\omega + i\gamma_c)}\ln\big[2\cosh(\mu_c/2k_B T)\big], \tag{1.36}$$

which coincides with the conductivity given by the classical Drude–Boltzmann approximation [25]. The real part of this term (zero for $\gamma_c = 0$) contributes to energy absorption or dissipation due to the intraband electrons. Except for a factor which accounts for the interlayer separation between graphene planes, equation (1.36) for $\mu_c = 0$ coincides with the expression derived by Wallace for graphite [11] and corresponds to the intraband conductivity of a single-wall carbon nanotube in the limit of infinite radius [12]. If $\mu_c \gg k_B T$, the carriers are degenerate and the material acts like a metal. In this case [30], $n_0 \approx \mu_c^2/\pi\hbar^2 v_F^2$, the chemical potential can be expressed as $\mu_c \approx E_F \approx \sqrt{\pi \hbar^2 v_F^2 n_0}$, where E_F is the Fermi energy, and the intraband term can be approximated by the Drude form

$$\sigma^{\text{intra}}(\omega) = \frac{ie^2|\mu_c|}{\pi \hbar^2(\omega + i\gamma_c)}. \tag{1.37}$$

It can be seen that $\text{Re}\,\sigma^{\text{intra}} \geqslant 0$ and $\text{Im}\,\sigma^{\text{intra}} > 0$, where Re and Im stand for the real and imaginary parts of a complex quantity. Therefore, σ^{intra} corresponds to a surface impedance Z (the inverse of the surface conductivity), with $\text{Im}\,Z < 0$, that is, an *inductive* surface impedance. As is well known for metallic boundaries [29], the sign of the imaginary part of the surface impedance plays an important role in the type of surface waves that can be guided by the boundary. Therefore, it is not surprising to find that, depending on the sign of the imaginary part of its complex surface conductivity, a graphene sheet can support TE or TM surface plasmons, a type of surface waves that represent the collective oscillations of surface charges [23, 30, 31], which will be discussed later.

1.8.2 Interband conductivity

The interband contribution to the conductivity also has real and imaginary parts. For low temperatures, $|\mu_c| \gg k_B T$, it can be written as [20, 25]

$$\sigma^{\text{inter}}(\omega) = \frac{e^2}{4\hbar}\left\{\frac{1}{2} + \frac{1}{\pi}\arctan\frac{\hbar(\omega + i\gamma_c) - 2\mu_c}{2k_B T}\right.$$
$$\left. - \frac{i}{2\pi}\ln\frac{\left[\hbar(\omega + i\gamma_c) + 2\mu_c\right]^2}{\left[\hbar(\omega + i\gamma_c) - 2\mu_c\right]^2 + (2k_B T)^2}\right\}, \tag{1.38}$$

which can be closely approximated by [13, 20, 22]

$$\sigma^{\text{inter}}(\omega) = \frac{e^2}{4\hbar}\left[\theta(\hbar\omega - 2|\mu|) + \frac{i}{\pi}\ln\left|\frac{\hbar(\omega + i\gamma_c) - 2|\mu_c|}{\hbar(\omega + i\gamma_c) + 2|\mu_c|}\right|\right], \tag{1.39}$$

where $\theta(\hbar\omega - 2|\mu|)$ is a step function. This expression shows that for $\gamma_c = 0$, $\text{Re}\,\sigma^{\text{inter}} = 0$ and $\text{Im}\,\sigma^{\text{inter}} < 0$, that is, the interband contribution to the conductivity is purely imaginary and negative. Therefore, under these conditions, σ^{inter} corresponds to a surface impedance Z with $\text{Im}\,Z > 0$, i.e. a *capacitive* surface impedance.

Other approximations for the interband contribution in different regimes can be found in the literature [25]. In the most general case, calculating the interband term requires that the integration indicated in the second term in equation (1.34) is performed. To avoid the singular integral, it is convenient to use the following equivalent form [25]

$$\sigma^{\text{inter}}(\omega) = \frac{e^2}{4\hbar}\left[G(\omega/2) - \frac{4(\omega + i\gamma_c)}{i\pi\hbar} \int_0^{+\infty} \frac{G(\epsilon) - G(\omega/2)}{(\omega + i\gamma_c)^2 - 4(\epsilon/\hbar)^2} d\epsilon \right], \qquad (1.40)$$

where $G(\epsilon) = f_0(-\epsilon) - f_0(\epsilon)$. Taking into account the fact that in the high-frequency limit, $\hbar\omega \gg (k_B T, \mu_c)$, $G(\epsilon) \to \tanh(\epsilon/2k_B T)$, this expression shows that, in this limit, $\sigma^{\text{inter}} \to 0$, and the total conductivity becomes mostly real and tends to the universal value $e^2/4\hbar$ [32], independent of any material parameters.

It is useful to write the conductivity in units of $e^2/\hbar$. In dimensionless form,

$$\sigma = \frac{e^2}{\hbar} \tilde{\sigma} = \alpha(4\pi\varepsilon_0 c) \tilde{\sigma}, \qquad (1.41)$$

with

$$\alpha = \frac{1}{4\pi\varepsilon_0} \frac{e^2}{\hbar c} = \frac{\mu_0}{4\pi} \frac{e^2 c}{\hbar} \qquad (1.42)$$

being the fine-structure constant. As a dimensionless quantity, α has the same numerical value, $\alpha^{-1} \approx 137.036$, in all systems of units. In the Gaussian unit system, where the elementary charge is measured in statcoulombs [1], instead of in coulombs as in the SI system, α takes the form $e^2/(\hbar c)$ and $\sigma_{(\text{Gaussian})} = \alpha c \tilde{\sigma}$.

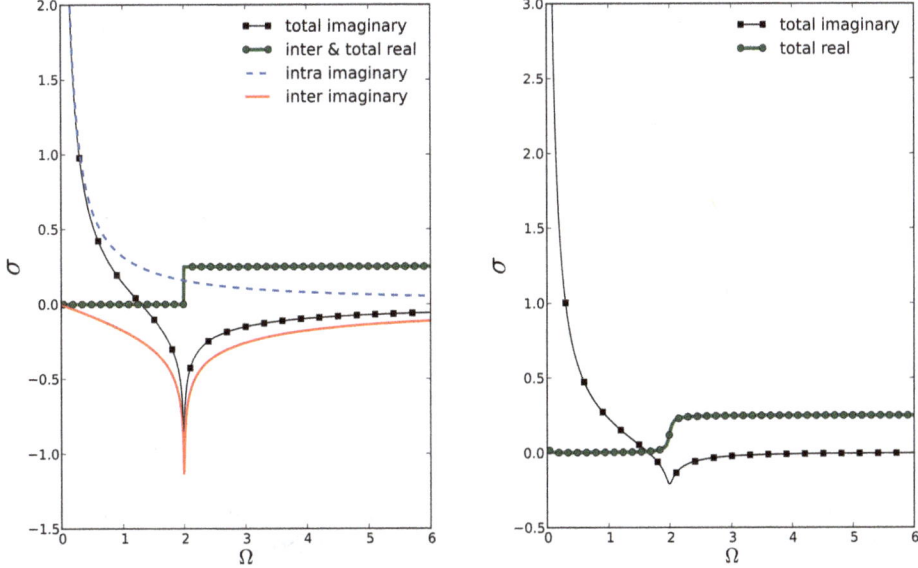

Figure 1.1. Imaginary and real parts of the surface conductivity of a graphene layer in units of $e^2/\hbar$ and as a function of $\Omega = \hbar\omega/\mu_c$. Left: Zero absolute temperature, collisionless limit, $\mu_c = 0.1$ eV, with intraband and interband contributions. Right: Non-zero temperature, $k_B T/\mu_c = 0.02$, $\mu_c = 0.9$ eV, $\gamma_c = 0.1$ meV. In the high-frequency limit the total conductivity becomes mostly real and tends to the universal value $e^2/4\hbar$. A Python script for obtaining the curves in this figure can be found in the appendix.

1.8.3 Material dispersion

In agreement with the experimental data [26–28], the Kubo model essentially predicts that: (i) interband contributions to the linear graphene surface conductivity are dominant in the frequency range corresponding to near-infrared and visible radiation, and (ii) intraband contributions are dominant, while interband contributions are negligible, in the frequency range corresponding to THz and far-infrared radiation.

The dispersive prototypical behavior of interband and intraband contributions is shown in figure 1.1 (left panel), corresponding to the zero absolute temperature collisionless limit ($\gamma_c = 0$), where σ depends on the combination $\hbar\omega/\mu$ (equations (1.37) and (1.39)). From these curves we see that when $\hbar\omega < 2\mu_c$, Re $\sigma = 0$, and thus conclude (see equation (1.33)) that no net conversion of electromagnetic energy into mechanical or heat energy is expected in a graphene sheet in this situation. On the other hand, when $\hbar\omega > 2\mu_c$, Re $\sigma > 0$, and this non-zero value is associated with energy absorption or dissipation in the graphene sheet. The dispersive behavior of the total surface conductivity for non-zero temperatures is shown in figure 1.1 (right panel) for $k_BT/\mu_c = 0.02$, a value for which the approximations involved in equations (1.36) and (1.38) apply (bear in mind that $k_BT \approx 0.03$ eV for an ambient temperature of 300 K).

We observe that the step-like singularity observed at low temperatures at $\Omega = \hbar\omega/\mu_c \approx 2$ and associated with the interband transition threshold can be spectrally tuned by changing the value of μ_c. For the value $\mu_c = 0.1$ eV used in figure 1.1(a), corresponding to a charge carrier density $n_0 \approx 8.14 \times 10^{11}$ cm^{-2}, the value $\Omega = 2$ occurs at a frequency equal to about 49 THz, whereas for $\mu_c = 0.9$ eV (figure 1.1(b), $n_0 \approx 6.6 \times 10^{13}$ cm^{-2}), the value $\Omega = 2$ occurs at about 436 THz. A Python script for obtaining the curves in this figure can be found in the appendix.

1.9 Decoupled equations

For linear, isotropic dielectric–magnetic media without external sources, Maxwell's equations in the frequency domain lead to the following decoupled second-order differential equations for the complex fields $E_\omega(r)$ and $H_\omega(r)$:

$$\frac{1}{\varepsilon(r, \omega)}\nabla \times \left[\frac{1}{\mu(r, \omega)}\nabla \times E_\omega(r)\right] = \frac{\omega^2}{c^2}E_\omega(r) \qquad (1.43)$$

$$\frac{1}{\mu(r, \omega)}\nabla \times \left[\frac{1}{\varepsilon(r, \omega)}\nabla \times H_\omega(r)\right] = \frac{\omega^2}{c^2}H_\omega(r), \qquad (1.44)$$

where, from now on and unless otherwise specified, ε is used to represent the effective dielectric constant $\varepsilon^{(\text{eff})}$ defined in equation (1.29).

When ε and μ are independent of position r (homogeneous media), the complex fields satisfy vector Helmholtz equations in the form

$$\nabla^2 F_\omega + \frac{\omega^2}{c^2}\varepsilon\mu F_\omega = 0, \qquad (1.45)$$

where F_ω stands for any of the phasors $E_\omega(r)$, $B_\omega(r)$, $D_\omega(r)$ and $H_\omega(r)$. A short calculation shows that *plane time harmonic* fields in the form

$$E_\omega(r) = a e^{ik\cdot r} \quad \text{and} \quad H_\omega(r) = \frac{c}{\mu\omega} (k \times a) e^{ik\cdot r} \tag{1.46}$$

satisfy equation (1.45) and are solutions of the homogeneous Maxwell equations provided $k \cdot a = 0$ (transversality condition, wave vector perpendicular to fields) and $k \cdot k = \omega^2 \varepsilon(\omega)\mu(\omega)/c^2$ (dispersion relation). The direction of the wave vector k may not generally coincide with the direction of power flow. The vector $\mathrm{Re}\, k$ points in the direction at which the phase of the wave propagates in space (the direction of the phase velocity). On the other hand, the direction in which the time-averaged power flow points (obtained from the real part of the complex Poynting vector [1]) is given by the real part of k/μ. It can be shown [33] that when the generally complex constitutive parameters of a linear isotropic dielectric–magnetic medium satisfy the condition

$$\mathrm{Re}\left\{ \frac{\varepsilon}{|\varepsilon|} + \frac{\mu}{|\mu|} \right\} > 0, \tag{1.47}$$

the phase of the wave propagates along the same direction as the electromagnetic power flow (positive phase velocity, or PPV, medium). Conversely, when this condition does not hold, the direction of the propagation of the phase of the wave is opposite to the direction of the electromagnetic power flow (negative phase velocity, or NPV, medium).

1.10 Boundary conditions

Although piecewise homogeneous materials are inhomogeneous, the inhomogeneities are completely confined to the boundaries. In this case it is convenient to formulate the solution of Maxwell's equations separately for each region, in which the constitutive parameters do not vary with position. At each position r, the set of differential equations (1.1)–(1.4) applies locally, except at the boundaries where partial derivatives only exist in the sense of distributions [34]. To obtain the solution for all space, *boundary condition* connecting solutions in separate regions must be specified. To do this, the integral equivalents of the differential version of Maxwell's equations are used in almost all textbooks [1].

To obtain the relations between the normal components of D and B on either side of the boundary between media 1 and 2, the integral equivalents of equations (1.3) and (1.4) are applied to the volume of a shallow Gaussian pillbox, half in medium 1 and half in medium 2. If the top and bottom pillbox faces are parallel and tangent to the boundary and n is a unit normal to the surface pointing from medium 1 to medium 2, we obtain [1]

$$(D^{(2)} - D^{(1)}) \cdot n = \rho^{(2D)} \tag{1.48}$$

$$(B^{(2)} - B^{(1)}) \cdot n = 0, \tag{1.49}$$

where $\rho^{(2D)}$, the surface charge density, is only different from zero when the external charges are confined exclusively to the interface.

Analogously, to obtain the relationships between the tangential components of E and H on either side of the boundary, the integral equivalents of equations (1.1) and (1.2) are applied to an infinitesimal Stokesian loop, with its long arms on either side of the boundary and its short arms perpendicular to the boundary. To describe the components of the fields that are tangent to the interface, two linearly independent directions are needed. If one direction is chosen along the normal t to the plane surface spanning the loop and the other along the unit vector $t \times n$, after application of Stoke's theorem we obtain [1]

$$n \times (E^{(2)} - E^{(1)}) = 0 \qquad (1.50)$$

$$n \times (H^{(2)} - H^{(1)}) = K, \qquad (1.51)$$

where it is understood that the surface current K only has components parallel to the surface.

In most practical examples, free charges and currents are distributed throughout a volume, consequently both $\rho^{(2D)}$ and K vanish. In problems that involve graphene, this is the case when graphene is modeled as a homogeneous film of finite thickness characterized by the effective dielectric constant equation (1.32). However, there are particular cases in which $\rho^{(2D)}$ and K do not vanish, for example at the surface of a perfect conductor. Most relevantly for our present purposes, $\rho^{(2D)}$ and K do not generally vanish at a graphene boundary when graphene is treated as an infinitely thin layer characterized by the constitutive equation (1.30). In the latter case and after turning to the frequency domain representation, equation (1.51) takes the form

$$n \times \left(H_\omega^{(2)} - H_\omega^{(1)} \right) = \sigma(\omega)\, E_\omega^{\parallel}, \qquad (1.52)$$

where $E_\omega^{\parallel}$ is the tangential component of E_ω and, according to equation (1.50), is continuous across the interface.

References

[1] Jackson J D 1998 *Classical Electrodynamics* 3rd edn (New York: Wiley)
[2] Weiglhofer W S 2003 *Introduction to Complex Mediums for Optics and Electromagnetics* ed W S Weiglhofer and A Lakhtakia (Bellingham, WA: SPIE) pp 27–61
[3] Mackay T G and Lakhtakia A 2010 *Electromagnetic Anisotropy and Bianisotropy: A Field Guide* (Singapore: World Scientific)
[4] Toll J S 1956 Causality and the dispersion relation: logical foundations *Phys. Rev.* **104** 1760–70
[5] Ponti S, Oldano C and Becchi M 2001 Bloch wave approach to the optics of crystals *Phys. Rev.* E **64** 021704
[6] Byron F W and Fuller R W 1992 *Mathematics of Classical and Quantum Physics* (New York: Dover) p 249
[7] Garg A 2012 *Classical Electromagnetism in a Nutshell* (Princeton, NJ: Princeton University Press)

[8] Sernelius B E 2012 Retarded interactions in graphene systems *Phys. Rev.* B **85** 195427

[9] Wolf E L 2014 *Applications of Graphene: An Overview* (New York: Springer)

[10] Novoselov K S, Geim A K, Morozov S V, Jiang D, Zhang Y, Dubonos S V, Grigorieva I V and Firsov A A 2004 Electric field effect in atomically thin carbon films *Science* **306** 666–9

[11] Wallace P R 1947 The band theory of graphite *Phys. Rev.* **71** 622–34
Wallace P R 1947 The band theory of graphite *Phys. Rev.* **72** 258 (erratum)

[12] Slepyan G Y, Maksimenko S A, Lakhtakia A, Yevtushenko O and Gusakov A V 1999 Electrodynamics of carbon nanotubes: dynamic conductivity, impedance boundary conditions, and surface wave propagation *Phys. Rev.* B **60** 17136

[13] Hanson G W 2008 Dyadic Green's functions and guided surface waves for a surface conductivity model of graphene *J. Appl. Phys.* **103** 064302

[14] Depine R A 1988 Scattering of a wave at a periodic boundary: analytical expression for the surface impedance *J. Opt. Soc. Am.* A **5** 507–10

[15] Sernelius B E 2012 Graphene as a strictly 2D sheet or as a film of small but finite thickness *Graphene* **1** 21–5

[16] Gusynin V P and Sharapov S G 2006 Transport of Dirac quasiparticles in graphene: Hall and optical conductivities *Phys. Rev.* B **73** 245411

[17] Gusynin V P, Sharapov S G and Carbotte J P 2006 Unusual microwave response of Dirac quasiparticles in graphene *Phys. Rev. Lett.* **96** 256802

[18] Peres N M R, Guinea F and Castro Neto A H 2006 Electronic properties of disordered two-dimensional carbon *Phys. Rev.* B **73** 125411

[19] Gusynin V P, Sharapov S G and Carbotte J P 2007 Magneto-optical conductivity in graphene *J. Phys.: Condens. Matter.* **19** 026222

[20] Gusynin V P, Sharapov S G and Carbotte J P 2007 Sum rules for the optical and Hall conductivity in graphene *Phys. Rev.* B **75** 165407

[21] Falkovsky L A and Pershoguba S S 2007 Optical far-infrared properties of a graphene monolayer and multilayer *Phys. Rev.* B **76** 153410

[22] Falkovsky L A and Varlamov A A 2007 Space–time dispersion of graphene conductivity *Eur. Phys. J.* B **56** 281–4

[23] Mikhailov S A and Ziegler K 2007 New electromagnetic mode in graphene *Phys. Rev. Lett.* **99** 016803

[24] Ziegler K 2007 Minimal conductivity of graphene: nonuniversal values from the Kubo formula *Phys. Rev.* B **75** 233407

[25] Falkovsky L A 2008 Optical properties of graphene and IV-VI semiconductors *Phys.–Usp.* **51** 887–97

[26] Dawlaty J M, Shivarman S, Strait J, George P, Chandrashekhar M, Rana F, Spencer M G, Veksler D and Chen Y 2008 Measurement of the optical absorption spectra of epitaxial graphene from terahertz to visible *Appl. Phys. Lett.* **93** 131905

[27] Li Z Q, Henriksen E A, Jiang Z, Hao Z, Martin M C, Kim P, Stormer H L and Basov D N 2008 Dirac charge dynamics in graphene by infrared spectroscopy *Nature Phys.* **4** 532–5

[28] Basov D N, Fogler M M, Lanzara A, Wang F and Zhang Y 2014 Colloquium: Graphene spectroscopy *Rev. Mod. Phys.* **86** 959–94

[29] Depine R A 1992 Backscattering enhancement of light and multiple scattering of surface waves at a randomly varying impedance plane *J. Opt. Soc. Am.* A **9** 609–18

[30] Luo X, Qiu T, Lu W and Zhenhua N 2013 Plasmons in graphene: recent progress and applications *Mater. Sci. Eng.* R **74** 351–76

[31] García de Abajo F J 2014 Graphene plasmonics: challenges and opportunities *ACS Photon* **1** 135–52

[32] Nair R R 2008 Fine structure constant defines visual transparency of graphene *Science* **320** 1308–1308

[33] Depine R A and Lakhtakia A 2004 A new condition to identify isotropic dielectric-magnetic materials displaying negative phase velocity *Microw. Opt. Technol. Lett.* **41** 315–6

[34] Idemen M M 2011 *Discontinuities in the Electromagnetic Field* (New York: Wiley)

Chapter 2

Single graphene sheet

This chapter is dedicated to the study of one of the most classical problems in macroscopic electrodynamics, that is, the problem of a single planar interface. The problem of the reflection and refraction of a wave at a planar interface between two dielectrics, also known as the Fresnel problem, can be found in many textbooks, e.g. ([1] pp 123ff, [2] pp 302–8). We begin this chapter on planar geometries by providing the detailed solution of the graphene-related variant of this problem, that is, when the planar interface that separates the two dielectrics is a graphene sheet. Compared with the problem without graphene, the main change in the general procedure resides in the boundary conditions, which are no longer expressed as the continuity across the interface of an unknown function and the continuity of a quantity combining the normal derivative of the unknown function and a constitutive parameter, but now include an impedance or Robin boundary condition [3], derived from equation (1.52), which can be considered as a weighted combination of the boundary conditions used in the problem without graphene. This type of boundary condition has been applied widely in the microwave range and in antenna theory and, although less common, is used occasionally [4] in studies of diffraction by metallic rough surfaces [5, 6].

The chapter proceeds with the homogeneous problem of a single planar graphene interface, a problem which is intimately related to the Fresnel problem. The homogeneous or mode problem of a graphene interface consists in finding the electromagnetic normal modes of this configuration, that is, the solutions to Maxwell's equations in the absence of external sources. The result is that doped graphene sustains localized normal modes called surface plasmon polaritons (SPPs) [7–13], represented by electromagnetic surface waves propagating along the graphene sheet and coupled to the plasma waves of the graphene two-dimensional (2D) electron gas. In contrast to metallic SPPs [14], graphene SPPs can be tuned through the application of an external voltage that is able to change the density of the

2-1

conduction carriers within a wide range. The plasmonic response of graphene is particularly strong in the THz frequency range at room temperature [15], whereas in conventional 2D electron gases, low temperatures are required to obtain plasmonic excitation. While metals support only p-polarized SPPs, graphene supports both s- and p-polarized SPPs [16, 17], one at high frequencies, where the imaginary part of graphene's conductivity is negative, and the other at low frequencies, where the imaginary part of graphene's conductivity is positive. This chapter reviews the theoretical description and the main characteristics of both types of graphene SPPs and it closes with a brief look at the numerical solution of the SPP dispersion equation.

2.1 Fresnel problem for a planar graphene sheet

We consider a graphene monolayer located at $y = 0$ separating two half-spaces with two homogeneous and isotropic materials characterized by the constitutive parameters ε_i (electric permittivity) and μ_i (magnetic permeability), $i = 1, 2$, see figure 2.1. Medium 1 ($y > 0$, the medium of incidence) is a conventional material with a positive refractive index $\nu_1 = \sqrt{\varepsilon_1\mu_1}$, $\varepsilon_1 > 0$, $\mu_1 > 0$, while medium 2 ($y < 0$, the medium of transmission) has the frequency-dependent constitutive parameters $\varepsilon_2 = \varepsilon_{2R} + i\varepsilon_{2I}$ and $\mu_2 = \mu_{2R} + i\mu_{2I}$, and a refractive index $\nu_2 = \pm\sqrt{\varepsilon_2\mu_2}$. It has been shown [18] that graphene can be used to build metamaterials and improve their characteristics. Therefore, to include engineered metamaterials with positive and negative refractive indices in the analysis, we let the real parts ε_{2R} and μ_{2R} have an arbitrary sign. The imaginary parts ε_{2I} and μ_{2I} must always be positive, independent of the signs of their real parts [19–21]. In the medium of transmission, the phase velocity and the group velocity are either parallel (they have the same sign, $\mathrm{Re}\,\nu_2 > 0$) or antiparallel (they have opposite signs, $\mathrm{Re}\,\nu_2 < 0$), depending on the constitutive parameters ε_2 and μ_2, respectively, satisfying or not satisfying condition equation (1.47).

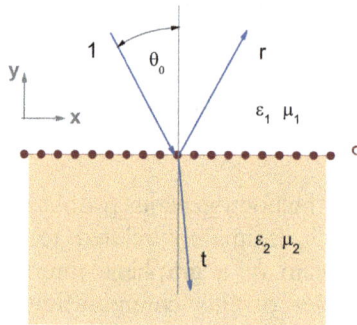

Figure 2.1. Reflection and transmission at a planar graphene sheet located at $y = 0$. A graphene monolayer characterized by a surface conductivity σ separates two half-spaces containing linear, homogeneous and isotropic materials. This boundary is illuminated by a linearly polarized incident electromagnetic wave with unit amplitude. In the s or TE polarization case the electric field of the incident wave is along the z direction, whereas in the p or TM polarization case the magnetic field of the incident wave is along the z direction. In both cases the reflected and transmitted fields conserve the polarization of the incident wave.

Graphene is treated as an infinitesimaly thin current sheet, with surface conductivity $\sigma(\omega)$, which is constant along the plane $y = 0$ (uniform doping, ideally no defects) and is given by the Kubo formula (equation (1.34)). This sheet is illuminated by the linearly polarized, time-harmonic (angular frequency ω) plane wave described by equation (1.46). This wave propagates in the (x, y) plane (incidence plane) and forms an angle θ_0 with the y axis. We analyze two independent polarization cases separately: the s or TE polarization (electric field in the z direction) and the p or TM polarization (magnetic field in the z direction). In both cases the reflected and transmitted fields conserve the polarization of the incident wave.

For a monochromatic incident wave, the phasor $F_\omega(r)$ in the integrand in equation (1.20) is proportional to a frequency Dirac delta function. Thus, to simplify notation, from now on we abandon the subscript ω in the Fourier transform of any time-dependent field component and use the same symbol both for the real time-dependent field component $F(r, t)$ and the complex space-dependent phasor $F_\omega(r)$. In what follows we use $f(x, y)$ to denote the phasor of the z-directed component of either the total electric field (s polarization) or the total magnetic field (p polarization).

In the region $y > 0$, $f(x, y)$ represents the sum of the incident wave (unit amplitude) and the reflected wave (amplitude r)

$$f_1(x, y) = e^{i(\kappa x - \beta_1 y)} + r e^{i(\kappa x + \beta_1 y)}, \tag{2.1}$$

while in the region $y < 0$

$$f_2(x, y) = t e^{i(\kappa x - \beta_2 y)} \tag{2.2}$$

represents the wave transmitted into medium 2 (amplitude t). The quantity $\kappa = k_0 \nu_1 \sin \theta_0$, $k_0 = \omega/c$, represents the x component of the incident wave vector, a quantity which is conserved because of the assumption of uniform doping. The components along y of the wave vectors in each medium are given by

$$\beta_j = \left(k_0^2 \varepsilon_j \mu_j - \kappa^2\right)^{1/2}, \quad j = 1, 2. \tag{2.3}$$

β_1 must be real and positive for the reflected field to move away from the surface into the half-space $y > 0$. In the ideal case of a completely transparent transmission medium (ε_2 and μ_2 real), the choice of the branches of the square roots in β_2 depends on the PPV or NPV character of medium 2: the positive branch for PPV and the negative branch for NPV media, in order for the power flow associated with the transmitted wave to move away from the surface towards $y \to \infty$. On the other hand, in the real case of lossy transmission media (Im $\varepsilon_2 > 0$, Im $\mu_2 > 0$), β_2 is always complex with a nonzero imaginary part, Im $\beta_2 > 0$, so that the transmitted field attenuates for $y \to -\infty$. Note that the condition Im $\beta_2 > 0$ automatically sets the sign of Re β_2, independently of the signs of ε_{2R} and of μ_{2R}, i.e. independently of the PPV or NPV character of the transmission medium.

Using the Poynting theorem and assuming that medium 2 is completely transparent, we obtain the fraction of the time-averaged incident power that is reflected from the boundary (the reflectance R) and the fraction that is transmitted through the boundary (the transmittance T) in terms of Fresnel reflection and the transmission coefficients for each polarization mode:

$$R = |r|^2 \qquad\qquad T = \frac{\eta_1}{\eta_2} |t|^2 \operatorname{Re} \beta_2/\beta_1, \qquad (2.4)$$

where $\eta = \mu$ for s polarization and $\eta = \varepsilon$ for p polarization. Energy balance requires that $R + T + A = 1$, where A is the fraction of the incident power absorbed by the graphene sheet.

To obtain the amplitudes of the reflected and transmitted waves, boundary value problems with conditions (1.50) and (1.52) at $y = 0$ are explicitly solved for each polarization case.

2.1.1 s polarization

For s polarization, the electric field in each zone is $E_j = f_j(x, y)z$, $j = 1, 2$. The magnetic field is obtained from Ampère's law (1.2):

$$H_j = \frac{i}{\omega\mu_0\mu_j} z \times \nabla f. \qquad (2.5)$$

In terms of $f(x, y)$, boundary conditions (1.50) and (1.52) at $y = 0$ become

$$f_1 = f_2 \qquad \text{and} \qquad \frac{1}{\mu_2} \frac{\partial f_2}{\partial y} - \frac{1}{\mu_1} \frac{\partial f_1}{\partial y} = i\omega\mu_0\sigma f_2. \qquad (2.6)$$

Introducing expressions (2.1) and (2.2), a system of equations is obtained:

$$r - t = -1 \qquad \text{and} \qquad \frac{\beta_1}{\mu_1} r + \left(\frac{\beta_2}{\mu_2} + \omega\mu_0\sigma \right) t = \frac{\beta_1}{\mu_1} \qquad (2.7)$$

and the complex amplitudes r and t result:

$$r^s = \frac{\beta_1/\mu_1 - \beta_2/\mu_2 - \omega\mu_0\sigma}{\beta_1/\mu_1 + \beta_2/\mu_2 + \omega\mu_0\sigma} \qquad (2.8)$$

$$t^s = \frac{2\beta_1/\mu_1}{\beta_1/\mu_1 + \beta_2/\mu_2 + \omega\mu_0\sigma}. \qquad (2.9)$$

2.1.2 p polarization

In this polarization, the magnetic field in each zone is given by $H_j = f_j(x, y)z$, $j = 1, 2$. The electric field is obtained using Faraday's law (1.1):

$$E_j = -\frac{i}{\omega\varepsilon_0\varepsilon_j} z \times \nabla f. \qquad (2.10)$$

The boundary conditions at $y = 0$ take the form

$$\frac{1}{\varepsilon_1}\frac{\partial f_1}{\partial y} = \frac{1}{\varepsilon_2}\frac{\partial f_2}{\partial y} \qquad \text{and} \qquad f_2 - f_1 = -i\frac{\sigma}{\omega\varepsilon_0}\frac{1}{\varepsilon_2}\frac{\partial f_2}{\partial y}. \qquad (2.11)$$

Introducing expressions (2.1) and (2.2) into these boundary conditions,

$$\frac{\beta_1}{\varepsilon_1}r + \frac{\beta_2}{\varepsilon_2}t = \frac{\beta_1}{\varepsilon_1} \qquad \text{and} \qquad -r + \left(1 + \frac{\sigma}{\omega\varepsilon_0}\frac{\beta_2}{\varepsilon_2}\right)t = 1, \qquad (2.12)$$

and solving this system, the plane wave reflection and transmission coefficients for p polarization result:

$$r^p = \frac{\varepsilon_2/\beta_2 - \varepsilon_1/\beta_1 + \sigma/\omega\varepsilon_0}{\varepsilon_2/\beta_2 + \varepsilon_1/\beta_1 + \sigma/\omega\varepsilon_0} \qquad (2.13)$$

$$t^p = \frac{2\varepsilon_2/\beta_2}{\varepsilon_2/\beta_2 + \varepsilon_1/\beta_1 + \sigma/\omega\varepsilon_0}. \qquad (2.14)$$

2.1.3 Optics of a graphene monolayer

The reflection coefficients given by equations (2.8) and (2.13) verify that $r^p(\theta_0 = 0) = -r^s(\theta_0 = 0)$, in agreement with the fact that there is no difference between s- and p polarization states at normal incidence. In addition, for $\tilde{\sigma} = 0$ the coefficients (2.8) and (2.9) (s polarization) and (2.13) and (2.14) (p polarization) reduce to the well-known Fresnel coefficients of a graphene-free interface [1, 2], whereas for $\tilde{\sigma} \to \infty$ $r^s = -r^p = -1$ and $t^s = t^p = 0$, they reduce to the Fresnel coefficients of a perfectly conducting surface.

To avoid difficulties with electromagnetic unit systems, it is useful to rewrite the Fresnel coefficients in terms of the dimensionless surface conductivity $\tilde{\sigma}$. Using equation (1.41), we obtain

$$r^s = \frac{\beta_1/\mu_1 - \beta_2/\mu_2 - 4\pi\omega a\tilde{\sigma}/c}{\beta_1/\mu_1 + \beta_2/\mu_2 + 4\pi\omega a\tilde{\sigma}/c} = r^s_{igj} \qquad (2.15)$$

$$t^s = \frac{2\beta_1/\mu_1}{\beta_1/\mu_1 + \beta_2/\mu_2 + 4\pi\omega a\tilde{\sigma}/c} = t^s_{igj} \qquad (2.16)$$

$$r^p = \frac{\varepsilon_2/\beta_2 - \varepsilon_1/\beta_1 + 4\pi ca\tilde{\sigma}/\omega}{\varepsilon_2/\beta_2 + \varepsilon_1/\beta_1 + 4\pi ca\tilde{\sigma}/\omega} = r^p_{igj} \qquad (2.17)$$

$$t^p = \frac{2\varepsilon_2/\beta_2}{\varepsilon_2/\beta_2 + \varepsilon_1/\beta_1 + 4\pi ca\tilde{\sigma}/\omega} = t^p_{igj}, \qquad (2.18)$$

where in the right-hand side we introduce an alternative notation with subscripts that make explicit reference to the media involved in the reflection and transmission at a surface, that is, the subscripts in coefficients r_{igj} and t_{igj} refer to waves impinging from the i side into a single graphene layer separating medium i from medium j. For future reference and for comparison, we write the Fresnel coefficients of a graphene-free interface in a form that is common to both polarization cases:

$$r_{ij} = \frac{\beta_i/\eta_i - \beta_j/\eta_j}{\beta_i/\eta_i + \beta_j/\eta_j} \qquad \text{and} \qquad t_{ij} = \frac{2\beta_i/\eta_i}{\beta_i/\eta_i + \beta_j/\eta_j}, \tag{2.19}$$

with $\eta_j = \varepsilon_j$ for p polarization or $\eta_j = \mu_j$ for s polarization.

When the transmission medium is completely transparent, the plane wave reflection and transmission coefficients can be written in terms of the angles of incidence θ_0 and refraction θ_t, $\sin\theta_t = \nu_2 \sin\theta_t/\nu_1$

$$r^s = \frac{\nu_1 \cos\theta_0/\mu_1 - \nu_2 \cos\theta_t/\mu_2 - 4\pi\alpha\tilde{\sigma}}{\nu_1 \cos\theta_0/\mu_1 + \nu_2 \cos\theta_t/\mu_2 + 4\pi\alpha\tilde{\sigma}} \tag{2.20}$$

$$t^s = \frac{2\nu_1 \cos\theta_0/\mu_1}{\nu_1 \cos\theta_0/\mu_1 + \nu_2 \cos\theta_t/\mu_2 + 4\pi\alpha\tilde{\sigma}} \tag{2.21}$$

$$r^p = \frac{\varepsilon_2/(\nu_2 \cos\theta_t) - \varepsilon_1/(\nu_1 \cos\theta_0) + 4\pi\alpha\tilde{\sigma}}{\varepsilon_2/(\nu_2 \cos\theta_t) + \varepsilon_1/(\nu_1 \cos\theta_0) + 4\pi\alpha\tilde{\sigma}} \tag{2.22}$$

$$t^p = \frac{2\varepsilon_2/(\nu_2 \cos\theta_t)}{\varepsilon_2/(\nu_2 \cos\theta_t) + \varepsilon_1/(\nu_1 \cos\theta_0) + 4\pi\alpha\tilde{\sigma}}. \tag{2.23}$$

Fresnel coefficients for the p polarization case are given in [22, 23], but these coefficients refer to the ratio between the amplitudes of the reflected and incident, and transmitted and incident, *tangential components of the electric field*, instead of the ratios between the amplitudes of the z components of the magnetic field, as considered here. When this difference is taken into account, it is easy to check that the coefficients given by equations (2.17) and (2.18) and those given in [22, 23] predict the same physical results. Expressions (2.15)–(2.18) for the planewave reflection and transmission coefficients also agree with those given in [24] for nonmagnetic media.

Very simple results are obtained for a graphene sheet immersed in a dielectric medium [23, 24]. By putting $\varepsilon_1 = \varepsilon_2 = \varepsilon$ and $\mu_1 = \mu_2 = \mu$, we obtain

$$r^s = \frac{-2\pi\alpha\tilde{\sigma}\sqrt{\mu/\varepsilon}}{\cos\theta_0 + 2\pi\alpha\tilde{\sigma}\sqrt{\mu/\varepsilon}} \tag{2.24}$$

$$t^s = \frac{\cos\theta_0}{\cos\theta_0 + 2\pi\alpha\tilde{\sigma}\sqrt{\mu/\varepsilon}} \tag{2.25}$$

$$r^p = \frac{2\pi\alpha\tilde{\sigma}\sqrt{\mu/\varepsilon}\cos\theta_0}{1 + 2\pi\alpha\tilde{\sigma}\sqrt{\mu/\varepsilon}\cos\theta_0} \tag{2.26}$$

$$t^p = \frac{1}{1 + 2\pi\alpha\tilde{\sigma}\sqrt{\mu/\varepsilon}\cos\theta_0}. \tag{2.27}$$

The normal incidence transmittance of free-standing graphene in vacuum as a function of $\hbar\omega$ is shown in figure 2.2 for three values of the ambient temperature. The Python script used to obtain this figure is given in the appendix. In the high-frequency range of the electromagnetic spectrum, where the surface conductivity tends to the universal value $e^2/4\hbar$ (i.e. the normalized conductivity $\tilde{\sigma}$ tends to the value 1/4, see equation (1.40) and figure 1.1), the Fresnel coefficients at normal incidence can be approximated by

$$r^p = -r^s \approx 2\pi\alpha\tilde{\sigma} \qquad t^p = t^s \approx 1 - 2\pi\alpha\tilde{\sigma}. \tag{2.28}$$

Therefore, the high-frequency normal incidence transmittance and reflectance of free-standing graphene in vacuum are

$$T = \frac{1}{(1 + \pi\alpha/2)^2} \approx 1 - \pi\alpha \approx 97.7\% \qquad R = (\pi\alpha/2)^2\, T < 0.1\%. \tag{2.29}$$

These are the values obtained when $\hbar\omega \gtrsim 2\mu_c$, as can be seen in figure 2.2, corresponding to $\mu_c = 0.5$ eV. At these frequencies, the total conductivity becomes

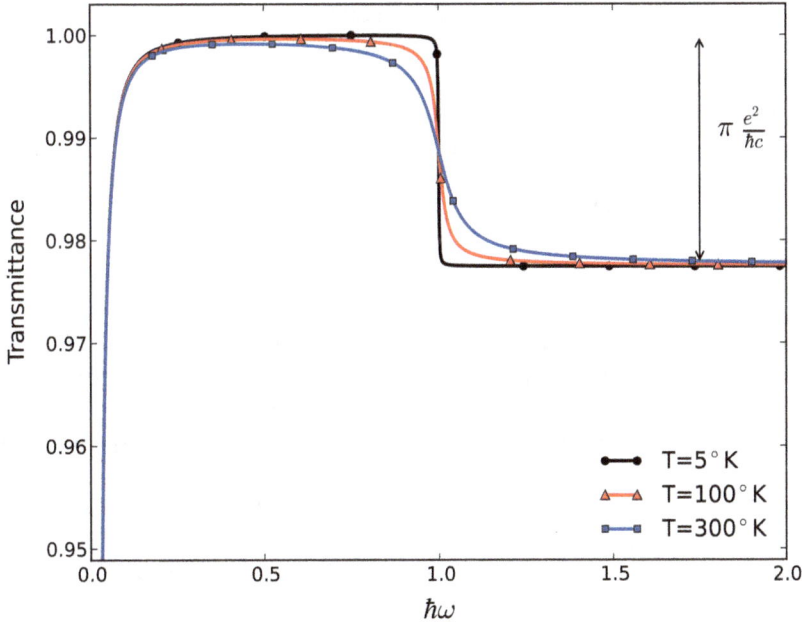

Figure 2.2. Normal incidence transmittance of free-standing graphene in vacuum as a function of $\hbar\omega$, $\mu_c = 0.5$ eV, $\hbar\gamma = 0.1$ meV.

mostly real and tends to the universal value $e^2/4\hbar$. This is the frequency range corresponding to $\Omega > 2$ in figure 1.1, where $\tilde{\sigma} \to 1/4$. The fraction of incident power that is absorbed by graphene results in $A \approx (1 - T) \approx \pi\alpha \approx 2.3\%$, a significant value, particularly considering that the graphene sheet is only one atom thick. These results for R and T were experimentally confirmed in [25]. We note that the optical observables R, A and T are defined by the fine structure constant α, a parameter which is not usually associated with material science, but rather with quantum electrodynamics. As noted in [25], the direct measurement of R and T in a stack of graphene planes is remarkable, not only because of the impressive agreement between experiment and theory, but also for providing evidence that the fine structure constant can be assessed directly, practically using the naked eye.

Equations (2.24)–(2.27) show that, in contrast to the well-known grazing incidence behavior at a material interface between two different dielectric media, where the interface reflects both s- and p-polarized waves almost totally, a self-standing graphene sheet immersed in a transparent medium behaves as a mirror for s-polarized waves but is almost transparent for p-polarized waves [24]. The predictions of equations (2.26) and (2.27) that $r^p(\theta_0 \to \pi/2) \to 0$ and $t^p(\theta_0 \to \pi/2) \to 1$ are in agreement with the fact that, at grazing incidence, the electric field of a p-polarized wave is perpendicular to the graphene sheet. Therefore, it is not able to induce electric currents and the total electromagnetic field remains unaltered and equal to that generated by the far sources (that is, to the the incident field). This behavior is illustrated in figure 2.3 with curves of the reflectance $R = |r^{s,p}(\theta_0)|^2$, transmitance $T = |t^{s,p}(\theta_0)|^2$ and graphene absorption $A = 1 - T - R$ as functions of the angle of incidence. The general trend of these curves is determined by the value of the frequency-dependent surface conductivity $\tilde{\sigma}(\omega)$ (see figure 1.1). To complement the results in the high-frequency $\Omega > 2$, where the total conductivity becomes mostly real and the approximations in equations (2.28) and (2.29) hold, in figure 2.3 we consider two frequency values in the low-frequency region $\Omega < 2$—(i) $\Omega = 0.016$, $\tilde{\sigma} = 0.17 + 20.38i$, $\lambda = 99.33$ μm, $\nu = 3.02$ THz and (ii) $\Omega = 0.1$, $\tilde{\sigma} = 0.007 + 3.18i$, $\lambda = 15.56$ μm, $\nu = 19.28$ THz—with conductivity values obtained from the Kubo model with $T = 300$ K, $\hbar\gamma_c = 0.1$ meV and $\mu_c = 0.8$ eV.

The behavior of the reflectance, transmittance and absorbance with an angle of incidence θ_0 is illustrated in figure 2.4 (see the appendix for the Python script) for the case when the graphene sheet separates two different materials ($\varepsilon_1 = 12$, $\varepsilon_2 = 4$, $\mu_1 = \mu_2 = 1$) and is illuminated from the more optically dense medium. The left column corresponds to $\tilde{\sigma} = 0.17 + 20.38i$ and the right one corresponds to $\tilde{\sigma} = 4.56 + 28.97i$. In contrast to the case considered in figure 2.3, the p polarization reflectance exhibits a minimum at an angle of incidence that is close to, but not equal to, the Brewster angle of the graphene-free boundary. The shift between the value of this quasi-Brewster angle and the Brewster angle of the boundary without graphene can be used to identify the presence of graphene in a sample [24]. The graphene layer does not change the position of the critical angle of total internal reflection, above which the transmitted wave becomes evanescent and the transmittance vanishes. However, in contrast to the graphene-free surface, above this angle the reflection is

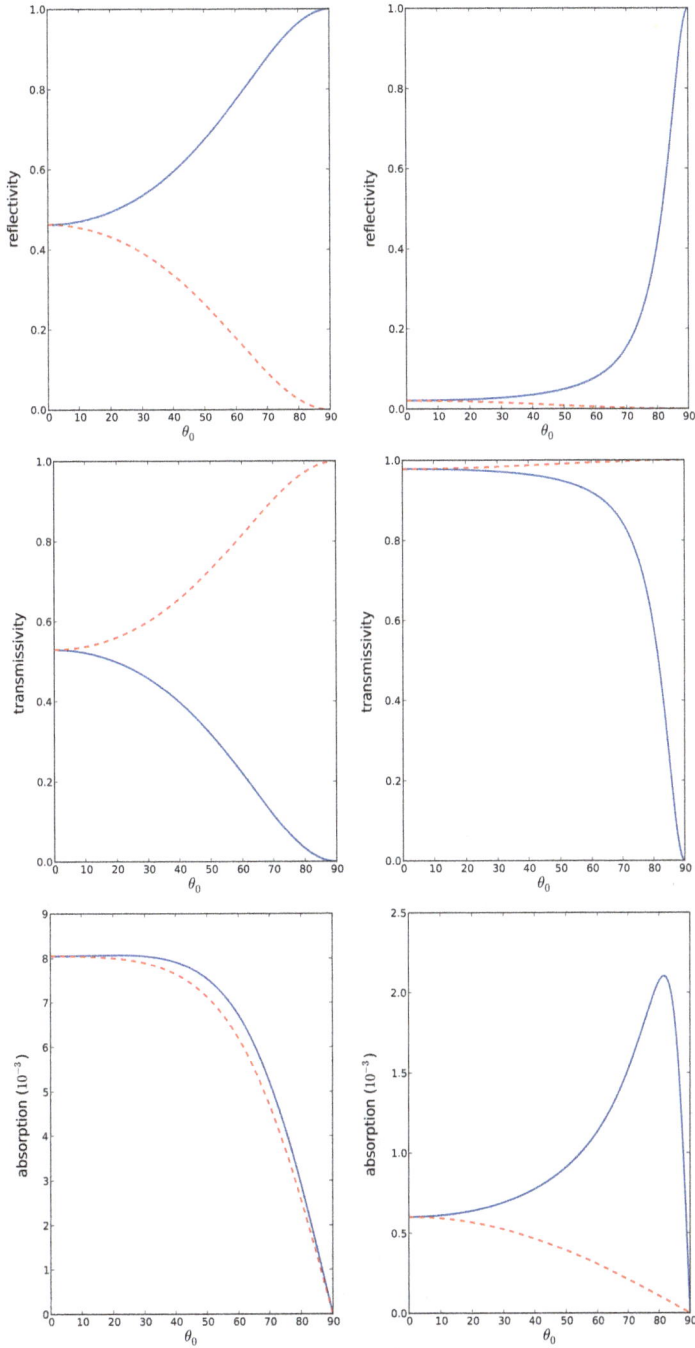

Figure 2.3. Curves of reflectance $R = |r^{s,p}(\theta_0)|^2$ (top row), transmitance $T = |t^{s,p}(\theta_0)|^2$ (middle row) and absorption $A = 1 - T - R$ (bottom row) as functions of the angle of incidence for a graphene sheet in air and for two frequency values in the region $\Omega < 2$: (i) left column, $\Omega = 0.016$, $\tilde{\sigma} = 0.17 + 20.38i$, $\nu = 3.02$ THz, and (ii) right column, $\Omega = 0.1$, $\tilde{\sigma} = 0.007 + 3.18i$, $\nu = 19.28$ THz. The Kubo parameters are $T = 300$ K, $\hbar\gamma_c = 0.1$ meV and $\mu_c = 0.8$ eV.

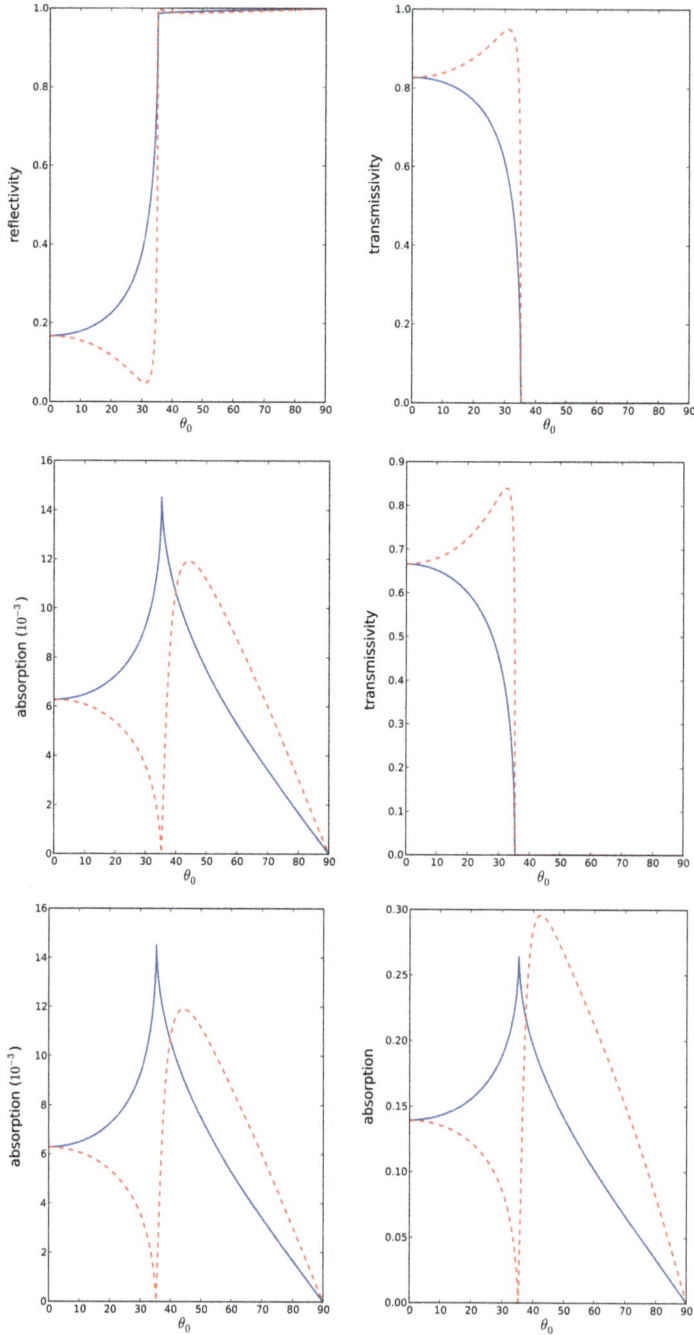

Figure 2.4. Curves of reflectance (top row), transmitance (middle row) and absorption (bottom row) as functions of the angle of incidence for a graphene sheet separating transparent media with $\varepsilon_1 = 12$, $\varepsilon_2 = 4$, $\mu_1 = \mu_2 = 1$. Left column: $\Omega = 0.016$, $\tilde{\sigma} = 0.17 + 20.38i$, $\nu = 3.02$ THz 99.33, $\mu_c = 0.8$ eV. Right column: $\Omega = 0.011$, $\tilde{\sigma} = 4.56 + 28.97i$, $\nu = 1.01$ THz (296.65), $\mu_c = 0.39$ eV. The other Kubo parameters are $T = 300$ K, $\hbar\gamma_c = 0.1$ meV. The Python script used to obtain these curves is given in the appendix.

not total, as can be better appreciated for the case with the higher value of $\mathrm{Re}\,\tilde{\sigma}$. In both of the cases considered in figure 2.4, $A^p \approx 0$, while A^s takes a maximum at an angle very close to the critical angle. A maximum also occurs in the curves of $A^p(\theta_0)$, but its position is shifted with respect to the angular position of the maximum of $A^s(\theta_0)$. For the case with the higher value of $\mathrm{Re}\,\tilde{\sigma}$ (right column), these maxima are $A^s \approx 0.26$ and $A^p \approx 0.3$.

2.2 The finite thickness model

The plane wave reflection and transmission coefficients given by equations (2.15)–(2.18) were obtained while assuming the validity of the infinitesimally thin current sheet model for graphene. However, there are occasions when it is preferable to abandon this idealization and model graphene as a homogeneous film of small but finite thickness δ, with an effective dielectric constant given by equation (1.32). For instance, this could be the case in problems where the finite thickness of the real atomic layer may influence the physics at close proximity, or when using numerical solvers which are not able to handle infinitesimally thin layers [26].

Because of its conceptual simplicity and because it is one of the most classical problems in macroscopic electrodynamics, the Fresnel problem serves as a good example for appreciating the similarities and differences between using the strictly 2D conductivity surface model and the finite thickness model to study electromagnetic problems involving graphene layers. Problems involving cylindrical and spherical geometries, where the graphene sheet can be treated as a homogeneous film that can be described by constant values of one of the coordinates, were discussed in [27]. In what follows we use the finite thickness model to obtain the plane wave reflection and transmission coefficients for a graphene monolayer separating two different media and show that when the thickness tends to zero, these results tend towards those obtained in the previous section using the Dirac delta model.

2.2.1 The problem for three regions

Instead of using the problem sketched in figure 2.1, we consider that the graphene sheet is replaced by a thin film of width δ located between $y = -\delta$ and $y = 0$ (see figure 2.5). In other words, we replace the problem with two regions and a single interface by another problem with three regions (medium 1, the film medium labeled with the index g and medium 2) and two interfaces. The expressions for the amplitudes of the s- and p-polarized waves reflected into medium 1 (r) and transmitted through the graphene layer into medium 2 (t) may be derived directly by applying, at $y = -\delta$ and $y = 0$, boundary conditions (1.50) and (1.51). Note that, due to the finite thickness δ, the currents inside the film are distributed throughout a volume and therefore $\mathbf{K} = 0$ in equation (1.51). Alternatively, the expressions for the amplitudes r and t in the presence of a homogeneous film can be obtained using Airy summation [28]. The result, valid for both s- and p polarization, is [29]

$$r = \frac{r_{1g} + r_{g2}e^{i2\beta_g\delta}}{1 + r_{1g}r_{g2}e^{i2\beta_g\delta}} \tag{2.30}$$

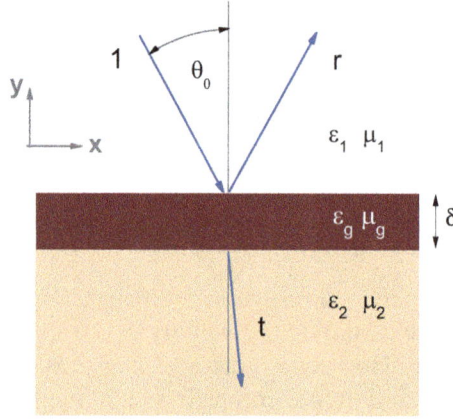

Figure 2.5. Graphene as a homogeneous film.

$$t = \frac{t_{1g}t_{g2}e^{i\beta_g\delta}}{1 + r_{1g}r_{g2}e^{i2\beta_g\delta}}, \tag{2.31}$$

where r_{ij} and t_{ij}, given by equation (2.19), are the amplitude reflection and transmission coefficients at an interface between medium i and j for waves impinging from the i side. Note that, in addition to the values 1 and 2 which make reference to media above and below the graphene sheet, i and j can also take the value g when making reference to the homogeneous medium corresponding to the thin film used to model graphene. The quantities β_j are still given by equation (2.3), with $\mu_g = 1$ and $\varepsilon_g = (1 + i\sigma/(\omega\varepsilon_0\delta))$, as it corresponds to a non-magnetic material with a complex permittivity given by equation (1.32).

2.2.2 The $\delta \rightarrow 0$ limit

To show the equivalence between the Dirac delta and finite thickness models, we insert the expressions given by equation (2.19) into equations (2.30) and (2.31), where ε_g and β_g now depend on δ, and we let the film thickness δ go towards zero. After some algebra, in this limit we arrive at the results in equations (2.8), (2.9), (2.13) and (2.14). For example, using the first order approximation for the exponential function in the amplitude reflection coefficient for s- and p polarization, equation (2.30) can be rewritten as

$$r = \frac{\dfrac{\beta_1}{\eta_1} - \dfrac{\beta_2}{\eta_2} + \left(\dfrac{\beta_1\beta_g}{\eta_1} + \dfrac{\beta_g^2}{\eta_g} - \dfrac{\beta_1\beta_2\eta_g}{\eta_1\eta_2} - \dfrac{\beta_g\beta_2}{\eta_2}\right)(i\delta + \mathcal{O}(\delta))}{\dfrac{\beta_1}{\eta_1} + \dfrac{\beta_2}{\eta_2} + \left(\dfrac{\beta_1\beta_g}{\eta_1} - \dfrac{\beta_1\beta_2\eta_g}{\eta_1\eta_2} - \dfrac{\beta_g^2}{\eta_g} + \dfrac{\beta_g\beta_2}{\eta_2}\right)(i\delta + \mathcal{O}(\delta))}. \tag{2.32}$$

Taking into account the fact that β_g depends on δ in the form $\beta_g = \sqrt{a + b/\delta}$, where a and b are constants independent of δ, it follows that $\beta_g^2\delta = a\delta + b$. And recalling that in the s polarization case $\eta_g = \mu_g = 1$, we obtain that in this case the amplitude reflection coefficient given by equation (2.30) coincides, in the limit $\delta \to 0$, with the expression given by equation (2.8). For the p polarization case a similar procedure can be used, except that now the quantity $\eta_g = \varepsilon_g$ also depends on δ, $\varepsilon_g \delta = \delta + i\sigma/(\omega \varepsilon_0)$ and therefore the amplitude reflection coefficient given by equation (2.30) coincides, in the limit $\delta \to 0$, with the expression given by equation (2.13).

Numerical simulations where the graphene monolayer is treated as a δ-thick layer with the equivalent complex effective permittivity $\varepsilon_g(\delta)$ have shown [30] that as long as the thickness is extremely small, the particular choice for the thickness value is not essential, a result that is theoretically justified by the limiting process shown in this section.

2.3 Graphene surface plasmon polaritons

The coherent oscillations of the delocalized electrons that exist at the interface between a conductor and a dielectric are known as surface plasmons. As the oscillation of surface charges produces electromagnetic fields in both the conductor and in the dielectric, the term surface plasmon polariton (SPP) waves [14, 31, 32] is used to indicate that the oscillation involves both surface-charge-density variations in the conductor ('surface plasmon') and electromagnetic waves in the air or dielectric ('polariton'). SPPs are a type of electromagnetic surface wave, i.e. they propagate along the interface while they are evanescently confined in the perpendicular direction. The localization at the interface is characterized by an enhancement of the local electromagnetic field (a characteristic which finds many applications, for example to boost light–matter interactions) and is highly sensitive towards changes in the geometry of the interface or the constitutive properties of the surrounding media (a characteristic which finds an obvious application in sensors, among many other areas).

The necessary condition for the existence of SPP waves on a flat surface between homogeneous media is that the media on both sides of the surface should have a constitutive parameter (electric permittivity or magnetic permeability) with opposite signs [33]. As a result of the fact that almost all natural materials have magnetic permeability with a positive real part, the most studied case in the literature is that of SPPs at interfaces between a dielectric and a metal or a plasma [14, 32, 34, 35], where opposite signs occur because of the negative electric permittivity of the metal/plasma. Much less attention has been paid to the case of SPPs at the boundary between a conventional dielectric and a metamaterial with a negative real part of its magnetic permeability, despite the fact that SPPs on metamaterial surfaces may have very different characteristics [36] from SPPs on metallic surfaces. The most significant differences are manifested in the polarization and direction of energy propagation. Whereas flat metallic surfaces can only support p polarized SPPs, flat metamaterial surfaces can support both p- and s-polarized SPPs, depending on the values taken by the constitutive parameters at the operating frequency. Analogously, in the metal case the direction of the energy propagation is always parallel to the direction of the wave propagation, while in the metamaterial case the direction of the energy propagation can be either parallel or antiparallel to the direction of the wave propagation [33].

In addition to noble metals, other plasmonic materials have been created that allow the dynamic manipulation of carrier densities, such as doped semiconductor nanocrystals [37]. Bearing in mind the conductivity surface model for graphene, it should not come as a surprise that a graphene monolayer behaves as a plasmonic 2D material in certain frequency ranges. In fact, in the search for plasmonic materials with lower losses and greater confinement of electromagnetic energy [38], the advent of graphene has led to many promising new opportunities for future plasmonic applications [11, 12, 39, 40].

Although the field profiles corresponding to collective surface charge oscillations at the boundary of a plasmonic three-dimensional (3D) material and those taking place at a graphene layer look the same, the physical systems are qualitatively different because the charge carriers in graphene are essentially frozen in the transverse dimension. As noted in [11], plasmons in graphene are 2D collective excitations, i.e. they are *real* plasmons, in contrast to *surface* plasmons at the boundary of 3D materials, and this fact leads to qualitatively different dispersions for the plasmonic modes in these two systems. In contrast to the usual metals, graphene can support either *s*- or *p*-polarized SPPs, depending on whether $\mathrm{Im}\,\sigma$ is positive or negative [41]. The existence of *s*-polarized SPPs, with a dispersion relation quite different from that of *p*-polarized SPPs, has been demonstrated experimentally [42].

2.3.1 Homogeneous problem for a graphene monolayer

In order to investigate the physical properties of graphene SPPs, we apply Maxwell's equations to the flat interface already considered in section 2.1. Taking into account the fact that SPPs are a type of electromagnetic surface mode that exists without external excitation, we must look for non-trivial boundary value problem solutions that are very similar to the one considered in section 2.1 when we were finding Fresnel reflection and transmission coefficients—but now in the absence of external sources. In other words, we must investigate if it is possible to have a guided mode along the graphene monolayer at $y = 0$ that propagates towards $x \to \infty$, vanishes when $y \to \pm\infty$ and is without an incident wave (homogeneous problem).

Mathematically, for each polarization we are looking for solutions such as those described by equations (2.1) and (2.2), except for the following differences.

(i) The absence of external excitation now requires the incident field (the first term in the right-hand side of equation (2.1)) to be zero. Without the incident field, the inhomogeneous system of linear equations obtained from the boundary conditions at $y = 0$ for the unknown amplitudes r and t becomes homogeneous (which explains the designation as a homogeneous problem). We see that a necessary condition for a non-trivial solution to this problem to exist is that the determinant of this homogeneous system of linear equations must equal zero, which means that, unlike the Fresnel reflection and transmission coefficients (2.15)–(2.18), these solutions r and t are not unique but have a linear dependence.

(ii) As there are no external sources, the value of κ is no longer fixed by the conservation of the x component of the incident wave vector. Instead, for each polarization mode, κ is determined by the boundary conditions at $y = 0$. It is clear that, if a solution exists, κ represents the propagation constant of the SPP propagating along x with frequency ω, with $2\pi/\mathrm{Re}\,\kappa$ being the spatial period (and with $1/\mathrm{Im}\,\kappa$ the propagation length) of the coherent oscillations of the surface-charge-density variations and their associated electromagnetic fields. According to the discussion in (i), the existence of nontrivial solutions requires that the value of κ be a complex zero of the determinant of the system of linear equations used to obtain the Fresnel reflection and transmission coefficients. In other words, κ must be a complex pole of the Fresnel amplitudes r and t, as can be seen from the fact that, according to Cramer's rule, the determinant of the system of linear equations is proportional to the denominator of r and t.

(iii) The constants β_j, equation (2.3), are always complex, with nonzero imaginary parts, even in the ideal case in which the half-spaces above and below the graphene layer are filled by completely transparent media ($\mathrm{Im}\,\varepsilon_j = 0$, $\mathrm{Im}\,\mu_j = 0$). The branches of the square roots in equation (2.3) must be chosen in order to have fields evanescently confined in the perpendicular direction. This is because the real part of graphene's surface conductivity, however small, is always greater than zero and therefore, to reflect the lossy character of the graphene layer, the propagation constant κ must also be complex. Then, β_j is also always complex with a nonzero imaginary part and, in order for the fields in both of the half-spaces to decay exponentially with distance from the interface, the condition

$$\mathrm{Im}\,\beta_j > 0 \quad j = 1, 2 \tag{2.33}$$

must hold, independently of the PPV or NPV character of the media filling the half-spaces above and below the graphene layer.

According to the discussion above, the phasor of the non-zero z component of the SPP electromagnetic field is obtained from the homogeneous version of equations (2.1) and (2.2)

$$\left.\begin{aligned} f_1(x, y) &= r e^{i(\kappa x + \beta_1 y)} & y > 0 \\ f_2(x, y) &= t e^{i(\kappa x - \beta_2 y)} & y < 0 \end{aligned}\right\}, \tag{2.34}$$

where κ is the unknown, generally complex, propagation constant of the SPP. The quantities β_j are analytical continuations in the κ complex plane of the quantities β_j defined in equation (2.3) and are subject to condition euqation (2.33). The complex amplitudes of the SPP, r and t, are solutions of the homogeneous systems of linear equations obtained from the boundary conditions at $y = 0$, namely, r^s and t^s are solutions of the homogeneous version of the linear system equation (2.7)

$$r^s - t^s = 0 \qquad \text{and} \qquad \frac{\beta_1}{\mu_1}r^s + \left(\frac{\beta_2}{\mu_2} + \omega\mu_0\sigma\right)t^s = 0, \tag{2.35}$$

whereas r^p and t^p are solutions of the homogeneous version of the linear system equation (2.12).

$$\frac{\beta_1}{\varepsilon_1}r^p + \frac{\beta_2}{\varepsilon_2}t^p = 0 \qquad \text{and} \qquad -r^p + \left(1 + \frac{\sigma}{\omega\varepsilon_0}\frac{\beta_2}{\varepsilon_2}\right)t^p = 0 \tag{2.36}$$

2.3.2 Dispersion equations for graphene SPPs

The value of κ for each polarization mode is obtained by looking for the complex poles of the Fresnel amplitudes. The dispersion relation for s polarization is (see the common denominators in equations (2.15) and (2.16))

$$\frac{\beta_1}{\mu_1} + \frac{\beta_2}{\mu_2} = -4\pi\frac{\omega}{c}\alpha\tilde{\sigma}, \tag{2.37}$$

whereas the dispersion relation for p polarization is (see the common denominators in equations (2.17) and (2.18))

$$\frac{\varepsilon_1}{\beta_1} + \frac{\varepsilon_2}{\beta_2} = -4\pi\frac{c}{\omega}\alpha\tilde{\sigma}. \tag{2.38}$$

These dispersion relations give the propagation constant κ as an implicit function of the angular frequency ω. Equations (2.37) and (2.38) agree with the dispersion relations obtained in [43] using dyadic Green functions represented as Sommerfeld integrals and equation (2.38) reduces to the result presented in [11] for nonmagnetic media.

To see that both s- and p-polarized SPPs can propagate in graphene, that the frequency domains of existence for p- and s-modes do not overlap, and that the kind of mode which can be supported is determined by the imaginary part of graphene's surface conductivity [11, 18, 39, 41, 43], we consider the ideal case in which the half-spaces above and below the graphene layer contain completely transparent conventional dielectrics, with real and positive constitutive parameters. According to condition (2.33), the imaginary part of the left-hand side of equation (2.37) must be positive, whereas the imaginary part of the right-hand side of the same equation is positive or negative, depending on the sign of the imaginary part of $\tilde{\sigma}$. Therefore, the existence of s-polarized SPPs requires Im $\tilde{\sigma} < 0$, a condition which occurs when the absolute value of the negative imaginary part of $\sigma^{\text{inter}}(\omega)$ is larger than the absolute value of the positive imaginary part of $\sigma^{\text{intra}}(\omega)$ (see figure 1.1). It can be shown [41] that, in the collisionless limit and when the absolute temperature is zero, solutions of equation (2.37) exist—and s-polarized SPPs are supported—in the spectral region

$1.667 < \hbar\omega/\mu_c < 2$, and with heavy damping in the region $\hbar\omega/\mu_c > 2$, where $\mathrm{Re}\,\tilde{\sigma} > 0$ due to the step-like behavior of $\mathrm{Im}\,\sigma^{\mathrm{intra}}(\omega)$.

A similar analysis holds for equation (2.38), except that in this case, and according to condition (2.33), the imaginary part of the left-hand side of the dispersion equation for p-modes must be negative and we conclude that for p-polarized SPPs to propagate, $\mathrm{Im}\,\tilde{\sigma} > 0$. In the collisionless zero temperature limit corresponding to figure 1.1a, this condition occurs when $0 < \hbar\omega/\mu_c < 1.667$ [41]. The value of $\hbar\omega/\mu_c$ at which $\mathrm{Re}\,\tilde{\sigma}$ becomes appreciable at finite temperatures is shifted slightly towards smaller values relative to the value $\hbar\omega/\mu_c = 2$ at zero temperature. In this case, a finite value of $\mathrm{Re}\,\tilde{\sigma}$ (see figure 1.1(b)) produces a very small damping for p-polarized SPPs and a heavy damping for s-polarized SPPs [18, 41].

2.3.3 Free-standing graphene

For a graphene sheet immersed in a medium of relative constitutive parameters ε and μ, the SPP dispersion relations take the following form

$$\sqrt{\frac{\omega^2}{c^2}\varepsilon\mu - \kappa^2} + 2\pi\frac{\omega}{c}\alpha\tilde{\sigma} = 0 \qquad s \text{ polarization} \qquad (2.39)$$

$$1 + 2\pi\frac{c}{\omega}\alpha\tilde{\sigma}\sqrt{\frac{\omega^2}{c^2}\varepsilon\mu - \kappa^2} = 0 \qquad p \text{ polarization}, \qquad (2.40)$$

and agree with the dispersion relations obtained in 1989 [44], which were not in the context of graphene studies, but in the more general context of the natural oscillations of conducting films. Solving these equations for the propagating constant κ, we obtain

$$\kappa = \frac{\omega}{c}\sqrt{\varepsilon\mu - (2\pi\alpha\tilde{\sigma})^2} \qquad \mathrm{Im}\,\tilde{\sigma} < 0 \qquad s \text{ polarization} \qquad (2.41)$$

$$\kappa = \frac{\omega}{c}\sqrt{\varepsilon\mu - 1/(2\pi\alpha\tilde{\sigma})^2} \qquad \mathrm{Im}\,\tilde{\sigma} > 0 \qquad p \text{ polarization.} \qquad (2.42)$$

It is clear from these expressions that the following is true.
  (i) If $\tilde{\sigma}$ is an almost imaginary number, then κ is an almost real complex quantity, a characteristic that is necessary to have a truly propagating surface wave, that is, a wave in which the spatial period along the surface $\lambda_{\mathrm{SPP}} = 2\pi/\mathrm{Re}\,\kappa$ (the plasmon wavelength) is much greater than the attenuation distance $1/\mathrm{Im}\,\kappa$;
  (ii) $\mathrm{Re}\,\kappa$ is always greater than $\omega/c = 2\pi/\lambda$, with λ being the wavelength of a photon with a frequency ω. Therefore, λ_{SPP}, the spatial periodicity associated with these graphene SPPs, is always less than the spatial periodicity that can be induced by a photon propagating on the half-spaces above and below the graphene layer (the periodicity along the surface provided by such a photon would range from λ for grazing incidence to infinity for normal

incidence). We conclude that an SPP on a single graphene layer cannot be excited by light of any frequency propagating in the space above or below the layer. This is not only true for SPPs propagating along a graphene sheet immersed in a homogeneous medium such as that considered to obtain equations (2.41) and (2.42), but also for SPPs propagating along a plane graphene sheet separating two isotropic media. This characteristic can be understood as a consequence of the evanescent confinement of the fields in directions normal to the interface, that is, a consequence of the requirement that the transverse wavenumbers β_j, $j = 1, 2$ must be almost imaginary complex quantities.

2.3.4 Solving dispersion relations

In the case of a flat graphene sheet separating two media characterized by different, and generally complex, constitutive parameters, a numerical method is needed to find the complex roots of the dispersion relations (2.37) and (2.38). In practice, sometimes it is possible to avoid the use of numerical root finding methods and to obtain the dispersion relations (2.37) and (2.38) in explicit form after simplifying the algebra, either by exploiting a symmetry, as in the case of equations (2.41) and (2.42) for free-standing graphene sheets, or by assuming some limiting behavior that provides a good approximation to actual behavior or that can serve as a good initial guess for the numerical solver. Here we group some particular situations where the dispersion relations can be solved easily without using a complex root finder for general nonlinear equations.

Nonretarded approximation for p *polarization*
Usually, the value of the propagation constant for *p*-polarized SPPs in the terahertz region is much much greater than the modulus of the wave vector of light propagating in the space above or below the graphene layer at the frequency ω. In this case, the nonretarded approximation $\mathrm{Re}\,\kappa \gg \omega/c$ can be used in equation (2.38), and then $\beta_1 \approx \beta_2 \approx i\kappa$, and the dispersion relation can be rewritten explicitly as

$$\kappa \approx i\frac{\omega}{c}\frac{\varepsilon_1 + \varepsilon_2}{4\pi\alpha\tilde{\sigma}}. \tag{2.43}$$

In figure 2.6 we compare the exact (retarded) values against the values obtained with equation (2.43) for a graphene layer in air, with Kubo parameters $T = 300$ K, $\mu_c = 0.8$ eV and $\hbar\gamma = 0.8$ meV and for frequencies $5 < \nu < 40$ THz. As shown in this figure, the approximation works quite well in this spectral range, for both the real and the imaginary parts of the propagation constant. Incidentally, note the rather large values, compared with the values attained by the metallic plasmons, taken by the real part of the propagation constant in this spectral range.

Noting that a frequency $\nu = 1$ THz corresponds to an energy of 4.1 meV, figure 1.1 shows that the dominant contribution to the surface conductivity for THz and far infrared frequencies comes from intraband electrons. Therefore, we can obtain

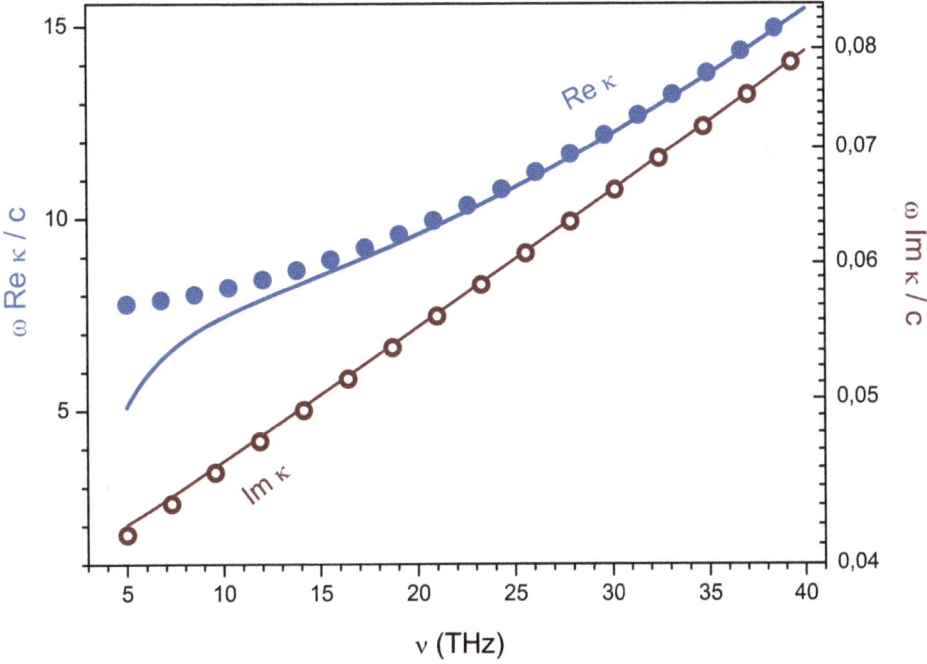

Figure 2.6. Real and imaginary parts of the propagation constant of p-polarized SPPs propagating along a graphene sheet in vaccum ($\varepsilon_1 = \mu_1 = \varepsilon_2 = \mu_2 = 1$). The surface conductivity obtained with the Kubo model with parameters $T = 300$ K, $\mu_c = 0.8$ eV and $\hbar\gamma = 0.8$ meV. Continuous curves: exact (retarded) values. Symbols: values obtained with the nonretarded approximation, equation (2.43).

simpler analytical results [13] by assuming that in this spectral region the graphene surface conductivity can be approximated by the Drude term given by equation (1.37). Doing this,

$$\tilde{\sigma}(\omega) = \frac{i|\mu_c|}{\pi(\hbar\omega + i\hbar\gamma_c)}, \tag{2.44}$$

and introducing this expression in the nonretarded dispersion relation (2.43), we obtain

$$\kappa \approx \frac{\omega}{c}\frac{\varepsilon_1 + \varepsilon_2}{4\alpha|\mu_c|}(\hbar\omega + i\hbar\gamma_c). \tag{2.45}$$

Thus, the dispersion of the real and imaginary parts of the propagation constant is given explicitly by

$$\mathrm{Re}\,\kappa \approx \frac{\omega}{c}\frac{\varepsilon_1 + \varepsilon_2}{4\alpha|\mu_c|}\hbar\omega \tag{2.46}$$

$$\mathrm{Im}\,\kappa \approx \frac{\omega}{c}\frac{\varepsilon_1 + \varepsilon_2}{4\alpha|\mu_c|}\hbar\gamma_c. \tag{2.47}$$

In the nonretarded approximation, both the period along the graphene sheet of the propagating surface plasmon and its penetration depth in the surrounding dielectrics are inversely proportional to $\mathrm{Re}\,\kappa$. Equation (2.46) shows that in this regime these characteristic lengths increase with the values of the frequency and the average dielectric constant, and decrease with the value of μ_c. For the parameters in figure 2.6 and for $\nu \approx 28$ THz (vacuum wavelength $\lambda \approx 10.7$ μm), $c\,\mathrm{Re}\,\kappa/\omega \approx 10$, a value which indicates that in this case both the graphene surface plasmon wavelength and the confinement of the fields in the immediate vicinity of the graphene sheet are ten times smaller than the photon wavelength.

Real and imaginary parts of κ for s polarization
To optimize the excitation and detection of s-polarized graphene surface plasmons it may be convenient [17] to have the real and imaginary parts of the SPP propagation constant κ as explicit functions on the real and imaginary parts of $\tilde{\sigma}$. The dispersion relation equation (2.41) provides an easy way to do so. Introducing $\kappa = \kappa_R + i\kappa_I$ and $\tilde{\sigma} = \tilde{\sigma}_R + i\tilde{\sigma}_I$ into this dispersion relation, after equating the real and imaginary parts we obtain

$$\left.\begin{array}{c} \kappa_R^2 - \kappa_I^2 - \dfrac{\omega^2}{c^2}\left[\varepsilon\mu + (2\pi\alpha)^2\left(\tilde{\sigma}_R^2 - \tilde{\sigma}_I^2\right)\right] = 0 \\[2em] \kappa_R\kappa_I = -\dfrac{\omega^2}{c^2}(2\pi\alpha)^2\,\tilde{\sigma}_R\tilde{\sigma}_I \end{array}\right\} . \tag{2.48}$$

Solving this system, we obtain κ_R^2

$$\begin{aligned} \kappa_R^2 = \dfrac{\omega^2}{c^2}\Bigg\{ &\left[\dfrac{\varepsilon\mu}{2} + 2\pi^2\alpha^2\left(\tilde{\sigma}_I^2 - \tilde{\sigma}_R^2\right)\right] \\ &+ \sqrt{\left[\dfrac{\varepsilon\mu}{2} + 2\pi^2\alpha^2\left(\tilde{\sigma}_I^2 - \tilde{\sigma}_R^2\right)\right]^2 + 2^4\pi^4\alpha^4\tilde{\sigma}_R^2\tilde{\sigma}_I^2}\Bigg\}, \end{aligned} \tag{2.49}$$

and κ_I is obtained by replacing this value in the second equation in equation (2.48). Considering the typical values for $\tilde{\sigma}$ shown in the examples given in figure 1.1, expression equation (2.48) shows that in most parts of the region of existence of s-polarized SPPs, $1.667 < \hbar\omega/\mu_c < 2$, the real part of the propagation constant takes values which are slightly greater than (but not very different from) the value $\omega\sqrt{\varepsilon\mu}/c$, corresponding to the modulus of the wave vector of light that propagates in the space above or below the graphene layer at the same frequency. This is in contrast to the behavior of the real part of the propagation constant of p-polarized SPPs, which, in their region of existence, $\hbar\omega/\mu_c < 1.667$, can take values that are very different from the wave vector of light propagating in the space above or below the graphene layer at the same frequency, as shown in the example given in figure 2.6.

Iterative method for p polarization
Whereas the propagation constant for p-polarized graphene plasmons propagating along a graphene sheet separating two identical half-spaces is explicitly given by the

simple analytical result in equation (2.40), a root-finding algorithm must be used to find the complex roots of the dispersion equation (2.38) in the general case where the graphene sheet separates two different media. For non-magnetic media, it is possible to use the following recursive numerical scheme, which avoids the use of root-finding algorithms. According to the general dispersion equation (2.38), the complex propagation constant $\kappa + \Delta\kappa$ for a p-polarized SPP propagating along a graphene sheet separating non-magnetic media with electric permittivities ε and $\varepsilon + \Delta\varepsilon$ satisfies the condition

$$\frac{\varepsilon + \Delta\varepsilon}{\sqrt{\frac{\omega^2}{c^2}(\varepsilon + \Delta\varepsilon) - (\kappa + \Delta\kappa)^2}} + \frac{\varepsilon}{\sqrt{\frac{\omega^2}{c^2}\varepsilon - (\kappa + \Delta\kappa)^2}} = -4\pi\frac{c}{\omega}\alpha\tilde{\sigma},$$
(2.50)

where

$$\kappa = \frac{\omega}{c}\sqrt{\varepsilon - 1/(2\pi\alpha\tilde{\sigma})^2}$$
(2.51)

is obtained from equation (2.42) and corresponds to the solution of equation (2.38) when $\Delta\varepsilon = 0$. Equation (2.50) can be rewritten in terms of the dimensionless complex propagation constant $q = c\kappa/\omega$ as

$$\frac{\varepsilon + \Delta\varepsilon}{\sqrt{(\varepsilon + \Delta\varepsilon) - (q + \Delta q)^2}} + \frac{\varepsilon}{\sqrt{\varepsilon - (q + \Delta q)^2}} = -4\pi\alpha\tilde{\sigma},$$
(2.52)

with

$$q = \sqrt{\varepsilon - 1/(2\pi\alpha\tilde{\sigma})^2}\,.$$
(2.53)

After a Taylor expansion of the the square roots in equation (2.50) in powers of Δq, and keeping only the first order term, we obtain

$$\frac{\varepsilon + \Delta\varepsilon}{R(\Delta\varepsilon,\, 0) - \Delta q\, q/R(\Delta\varepsilon,\, 0)} + \frac{\varepsilon}{R(0,\, 0) - \Delta q\, q/R(0,\, 0)} \approx -4\pi\alpha\tilde{\sigma},$$
(2.54)

where $R(\Delta\varepsilon,\, \Delta q) = [(\varepsilon + \Delta\varepsilon) - (\kappa + \Delta\kappa)^2]^{1/2}$. Keeping only first powers of Δq and after a little algebra, the relation

$$\Delta q = \frac{1}{q}\frac{N_1}{N_2}$$
(2.55)

is obtained, where
$$N_1 = (\varepsilon + \Delta\varepsilon)R(0,\, 0) + \varepsilon R(\Delta\varepsilon,\, 0) + 4\pi\alpha\tilde{\sigma}R(\Delta\varepsilon,\, 0)\, R(0,\, 0)$$

and

$$N_2 = \frac{\varepsilon + \Delta\varepsilon}{R(0,\, 0)} + \frac{\varepsilon}{R(\Delta\varepsilon,\, 0)} + 4\pi\alpha\tilde{\sigma}\left[\frac{R(0,\, 0)}{R(\Delta\varepsilon,\, 0)} + \frac{R(\Delta\varepsilon,\, 0)}{R(0,\, 0)}\right].$$

These expressions can be used in a recursive numerical scheme to solve the general dispersion equation (2.38) for p-polarized SPPs propagating along graphene surrounded by non-magnetic media with electric permittivities $\varepsilon_1 \neq \varepsilon_2$ m, starting from the particular solution given by equation (2.42) and corresponding to a

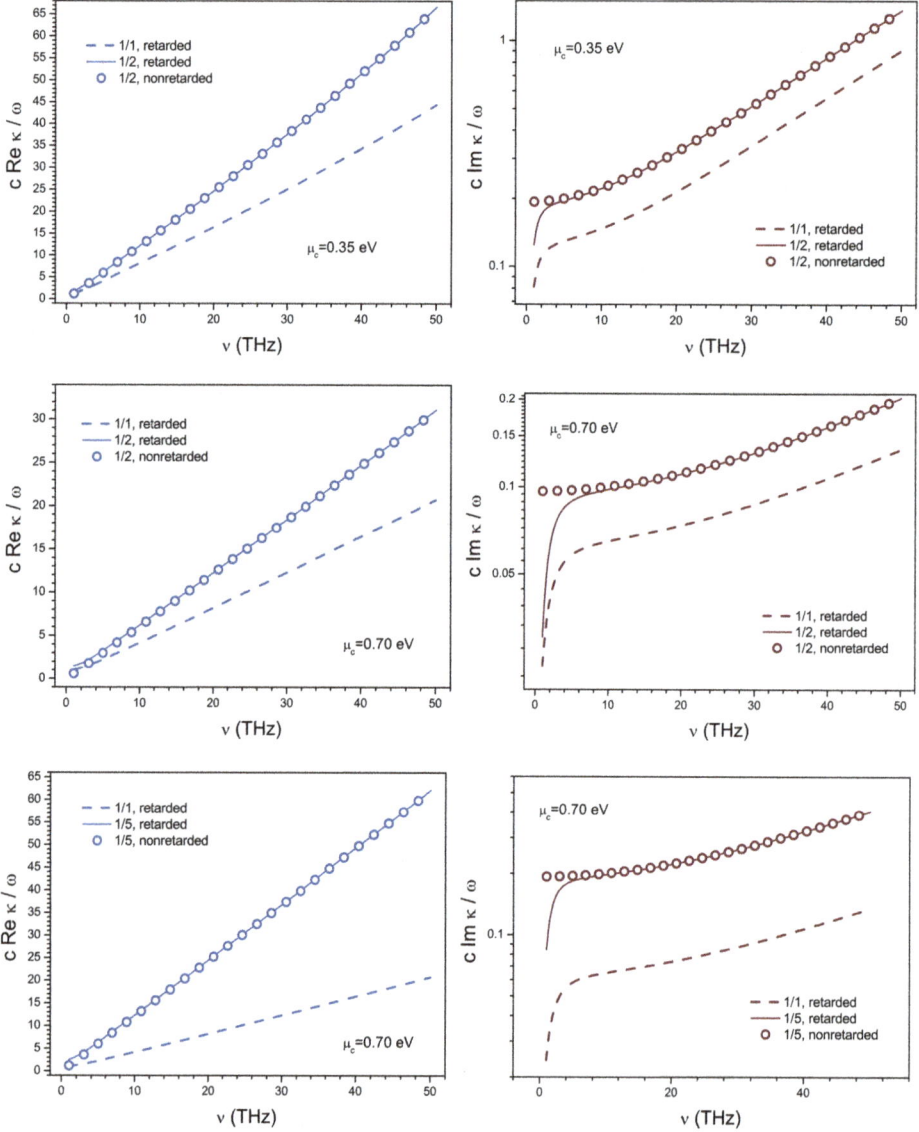

Figure 2.7. Real (left column) and imaginary (right column) parts of the propagation constant for p-polarized graphene SPPs as functions of the frequency in the spectral range $1 < \nu < 50$ THz. The continuous curves correspond to the configurations $\varepsilon_1/\varepsilon_2 = 2$, $\mu_c = 0.35$ eV (first row), $\varepsilon_1/\varepsilon_2 = 2$, $\mu_c = 0.70$ eV (second row) and $\varepsilon_1/\varepsilon_2 = 5$, $\mu_c = 0.70$ eV (third row). The values $T = 300$ K and $\hbar\gamma_c = 0.658$ meV were used as the Kubo parameters for all the curves. The circles correspond to the values obtained with the nonretarded approximation and the dashed curves correspond to the propagation constants for a free-standing graphene sheet immersed in vacuum. The Python script used to obtain these curves is given in the appendix.

graphene sheet surrounded by media with equal electric permittivity values. Assuming, for instance, that $\varepsilon_1 < \varepsilon_2$, the expression equation (2.55) allows us to obtain the propagation constant $\kappa + \Delta\kappa$ for the perturbed configuration ε_1–graphene– $\varepsilon_1 + \Delta\varepsilon$ from the value κ, corresponding to the unperturbed configuration ε_1–graphene–ε_1. With the help of a numerical parameter N, $\Delta\varepsilon = (\varepsilon_2 - \varepsilon_1)/N$ is conveniently chosen to minimize the error committed in the approximations that lead to equation (2.55). If this error is below a given tolerance, the scheme is repeated one more time, but this time considering that the unperturbed configuration corresponds to ε_1–graphene–$\varepsilon_1 + \Delta\varepsilon$ and that the perturbed configuration corresponds to ε_1–graphene– $\varepsilon_1 + 2\Delta\varepsilon$. If the absolute value of the difference between the left- and right-hand sides of the dispersion equation (2.38) is below the given tolerance at every step, we finally apply the scheme to the calculation of the propagation constant for the perturbed configuration ε_1–graphene–ε_2 in terms of the value for the unperturbed configuration ε_1–graphene– $\varepsilon_2 - \Delta\varepsilon$. If the error in a particular step is not below the given tolerance, the scheme can be repeated using a smaller value for $\Delta\varepsilon$.

We used this iterative scheme to find the dispersion relation for p-polarized graphene SPPs in different conditions (a Python script based on this scheme is given in the appendix). To illustrate the performance of the scheme and other numerical aspects described in this section, in figure 2.7 we show the real and imaginary parts of the SPP propagation constant as functions of the frequency in the spectral range $1 < \nu < 50$ THz (vacuum wavelengths between 300 and 6 μm) for the configurations $\varepsilon_1/\varepsilon_2 = 2$ and $\varepsilon_1/\varepsilon_2 = 4$, $T = 300$ K, $\hbar\gamma_c = 0.658$ meV and the values $\mu_c = 0.35$ eV and 0.7 eV. For these calculations, a tolerance value of 10^{-9} was used to control the fulfillment of condition equation (2.38). The nonretarded values obtained with expression (2.43) and the values corresponding to a free-standing graphene sheet immersed in vacuum are also given for comparison. The nonretarded approximation works quite well over almost all the frequency range considered in these figures, except for the lowest frequency values, for which the assumption that the real part of the propagation constant is much greater than the modulus of the wave vector of light propagating in the space above or below the graphene layer does not hold. However, above frequencies around 5 THz the approximation clearly holds in these cases, and the plasmon propagation constants take values that are unattainable by metallic plasmons.

References

[1] Chen H C 1983 *Theory of Electromagnetic Waves: A Coordinate Free Approach* (New York: McGraw-Hill)

[2] Jackson J D 1998 *Classical Electrodynamics* 3rd edn (New York: Wiley)

[3] Morse P M and Feshbach H 1953 *Methods of Theoretical Physics* (New York: McGraw-Hill) ch 6

[4] Yuferev S V and Ida N 2009 *Surface Impedance Boundary Conditions: A Comprehensive Approach* (Boca Raton, FL: CRC)

[5] Depine R A and Simon J M 1983 Surface impedance boundary condition for metallic diffraction gratings in the optical and infrared range *J. Mod. Opt.* **30** 313–22

[6] Depine R A 1988 Scattering of a wave at a periodic boundary: analytical expression for the surface impedance *J. Opt. Soc. Am.* A **5** 507–10

[7] Shung K W K 1986 Dielectric function and plasmon structure of stage-1 intercalated graphite *Phy. Rev.* B **34** 979–93

[8] Hwang E H and Das Sarma S 2007 Dielectric function, screening, and plasmons in two-dimensional graphene *Phy. Rev.* B **75** 205418

[9] Wunsch B, Stauber T, Sols F and Guinea F 2006 Dynamical polarization of graphene at finite doping *New J. Phys.* **8** 318

[10] Stauber T, Schliemann J and Peres N M R 2010 Dynamical polarizability of graphene beyond the Dirac cone approximation *Phy. Rev.* B **81** 085409

[11] Jablan M, Buljan H and Soljacic M 2009 Plasmonics in graphene at infrared frequencies *Phy. Rev.* B **80** 245435

[12] Kaminer I, Katan Y T, Buljan H, Shen Y, Ilic O, López J J, Wong L J, Joannopoulos J D and Soljacic M 2016 Efficient plasmonic emission by the quantum Cerenkov effect from hot carriers in graphene *Nature Commun.* **7** 11880

[13] Bludov Y V, Ferreira A, Peres N M R and Vasilevskiy M I 2013 A primer on surface plasmon-polaritons in graphene *Int. J. Mod. Phys.* B **27** 1341001

[14] Boardman A D (ed) 1982 *Electromagnetic Surface Modes* (New York: Wiley)

[15] Ju L *et al* 2011 Graphene plasmonics for tunable terahertz metamaterials *Nature Nanotech* **6** 630–4

[16] Jablan M, Buljan H and Soljacic M 2011 Transverse electric plasmons in bilayer graphene *Opt. Express* **19** 11236

[17] Mason D R, Menabde S G and Park N 2014 Unusual Otto excitation dynamics and enhanced coupling of light to TE plasmons in graphene *Opt. Express* **22** 847–58

[18] Luo X, Qiu T, Lu W and Zhenhua N 2013 Plasmons in graphene: recent progress and applications *Mater. Sci. Eng.* R **74** 351–76

[19] Landau L D, Lifshitz E M and Pitaevskii L P 1984 *Electrodynamics of Continuous Media* 2nd edn (New York: Wiley)

[20] Depine R A and Lakhtakia A 2004 Comment I on 'Resonant and antiresonant frequency dependence of the effective parameters of metamaterials' *Phys. Rev.* E **70** 048601

[21] Depine R A and Lakhtakia A 2015 Comment I on 'Poynting theorem constraints on the signs of the imaginary parts of the electromagnetic constitutive parameters' *J. Opt. Soc. Am.* A **32** 1564–5

[22] Falkovsky L A and Pershoguba S S 2007 Optical far-infrared properties of a graphene monolayer and multilayer *Phys. Rev.* B **76** 153410

[23] Falkovsky L A 2008 Optical properties of graphene and IV-VI semiconductors *Phys.-Usp* **51** 887–97

[24] Bludov Y V, Peres N M R and Vasilevskiy M I 2013 Unusual reflection of electromagnetic radiation from a stack of graphene layers at oblique incidence *J. Opt.* **15** 114004

[25] Nair R R 2008 Fine structure constant defines visual transparency of graphene *Science* **320** 1308–1308

[26] Sernelius B E 2012 Graphene as a strictly 2D sheet or as a film of small but finite thickness *Graphene* **1** 21–5

[27] Sernelius B E 2014 Electromagnetic normal modes and Casimir effects in layered structures *Phys. Rev.* B **90** 155457

[28] Airy G B 1833 On the phaenomena of Newton's rings when formed between two transparent substances of different refractive powers *Phil. Mag.* **2** 20–30

[29] Born M and Wolf E 1986 *Principles of Optics* 6th edn (Oxford: Pergamon)

[30] Vakil A and Engheta N 2011 Transformation optics using graphene *Science* **332** 1291–4

[31] Novotny L and Hecht B 2012 *Principles of Nano-Optics* 2nd edn (Cambridge: Cambridge University Press)

[32] Maier S A 2007 *Plasmonics: Fundamentals and Applications* (Berlin: Springer)

[33] Darmanyan S A, Nevière M and Zakhidov A A 2003 Surface modes at the interface of conventional and left-handed media *Opt. Commun.* **225** 233–40

[34] Agranovich V M and Mills D L (ed) 1982 *Surface Polaritons: Electromagnetic Waves at Surfaces and Interfaces* (Amsterdam: North-Holland)

[35] Raether H 1988 *Surface Plasmons on Smooth, Rough Surfaces and on Gratings* (Berlin: Springer)

[36] Cuevas M and Depine R A 2009 Radiation characteristics of electromagnetic eigenmodes at the corrugated interface of a left-handed material *Phys. Rev. Lett.* **103** 097401

[37] Schimpf A, Thakkar N, Gunthardt C, Masiello D and Gamelin D 2014 Charge-tunable quantum plasmons in colloidal semiconductor nanocrystals *ACS Nano* **8** 1065–72

[38] West P, Ishii S, Naik G, Emani N, Shalaev V and Boltasseva A 2010 Searching for better plasmonic materials *Laser Photon. Rev.* **4** 795–808

[39] Jablan J, Soljacic M and Buljan H 2013 Plasmons in graphene: fundamental properties and potential applications *Proc. of the IEEE* **101** 1689–704

[40] Rana F 2008 Graphene terahertz plasmon oscillators *IEEE Trans. Nanotechnol.* **7** 91–9

[41] Mikhailov S A and Ziegler K 2007 New electromagnetic mode in graphene *Phys. Rev. Lett.* **99** 016803

[42] Qiaoliang B, Han Z, Bing W, Zhenhua N, Candy Haley Y X L, Yu W, Ding Y T and Kian P L 2011 Broadband graphene polarizer *Nat. Photon.* **5** 411–5

[43] Hanson G W 2008 Dyadic Green functions and guided surface waves for a surface conductivity model of graphene *J. Appl. Phys.* **103** 064302

[44] Fal'ko V I and Khmel'nitskiĭ D E 1989 What if a film conductivity exceeds the speed of light? *Zh. Eksp. Teor. Fiz.* **95** 1988
Fal'ko V I and Khmel'nitskiĭ D E 1989 *Sov. Phys.-JETP* **68** 1150

IOP Concise Physics

Graphene Optics: Electromagnetic Solution of Canonical Problems

Ricardo A Depine

Chapter 3

Layered structures

After reviewing the general procedures for finding the electromagnetic response and modal fields of a planar graphene sheet separating two homogeneous half-spaces, in this chapter we demonstrate how these characteristics may be tuned by introducing additional layers of graphene. As discussed in section 2.3.3, and as is obvious from the explicit expressions (2.41) and (2.42) for the particular case of self-standing graphene sheets, the propagation constant κ of graphene SPPs in the general case described by equation (2.34) is always greater than the modulus of the wave vector of light propagating in the space above or below the graphene layer at the frequency ω. Therefore, in common with SPPs on a flat metal/dielectric interface, SPPs on a flat dielectric/graphene/dielectric interface cannot be excited directly by light beams unless special techniques for phase-matching are employed.

One of the most common techniques available for the excitation of metal or graphene SPPs relies on using multilayer films. This chapter begins with an introductory discussion of the prism coupler, that is, the simplest case of the multilayer coupler. Another phase-matching technique, the grating coupler, will be discussed in chapter 4. Then, the chapter reviews the general formalism for finding the electromagnetic response of a prism coupler involving a graphene sheet and gives examples in which this coupler is used for exciting s- and p-polarized graphene SPPs. Moving on to more general layered structures, the chapter closes with an analysis of the transmission properties of periodic stacks of parallel graphene sheets (one-dimensional (1D) photonic crystal).

3.1 Multilayers as couplers

The fact that the propagation constant κ of graphene SPPs at a single flat interface is always greater than the modulus of the wave vector of light propagating in the space above or below the graphene sheet at the frequency ω is a consequence of the requirement that the transverse wavenumbers β_j ($j = 1, 2$) must be almost imaginary complex quantities for the fields represented by equation (2.34) to be confined to the graphene sheet. This is clearly seen in the ideal case of lossless graphene and completely transparent media, where condition (2.33) leads to purely real values for κ, satisfying the relation $\kappa > \omega\sqrt{\varepsilon_j\mu_j}/c$. In the real case, graphene exhibits dissipation losses, κ becomes complex, and it is the real part of κ that is greater than the modulus of the wave vector of light in both media.

The frequency–wavenumber diagram in figure 3.1 illustrates the dispersion relation for p-polarized SPPs in the spectral range between 1.5 and 20 THz for an ideal graphene sheet separating two ideal non-dispersive dielectrics with constitutive parameters $\varepsilon_1 = \mu_1 = 1$ and $\varepsilon_2 = 2$, $\mu_2 = 1$. The curve for $\omega(\kappa)$ was obtained by numerically solving equation (2.38) after neglecting the small real part of $\bar{\sigma}$ obtained with Kubo parameters $T = 300$ K, $\mu_c = 0.8$ eV and $\hbar\gamma = 0.1$ meV. Note that the curve $\omega(\kappa)$ is always at the right of the gray vertical lines located at $c\kappa/\omega = \sqrt{\varepsilon_1\mu_1}$ and $c\kappa/\omega = \sqrt{\varepsilon_2\mu_2}$, that is, at the right of the values corresponding to the wave-vectors of light propagating on both sides of the graphene sheet.

The sketch in figure 3.2(a) illustrates the attempt to excite SPPs on a flat dielectric/graphene/dielectric interface using incident light from medium 1. The semicircle in medium 1 represents the dispersion relation for plane waves in the medium of incidence. As the magnitude of the propagation constant κ is greater than the radius

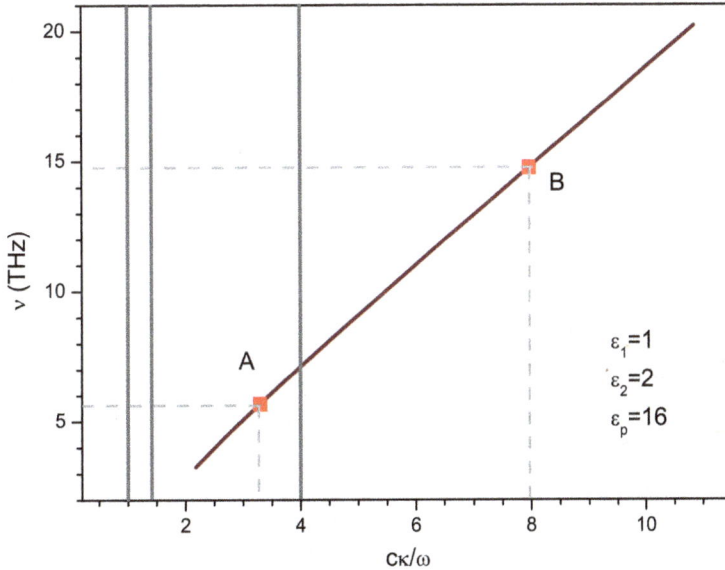

Figure 3.1. Dispersion relation for p-polarized SPPs corresponding to an ideal graphene sheet separating medium 1 ($\varepsilon_1 = \mu_1 = 1$) and medium 2 ($\varepsilon_2 = 2$, $\mu_2 = 1$). $\kappa(\omega)$ was obtained by numerically solving equation (2.38) after neglecting the small real part of $\bar{\sigma}$ given by the Kubo model for $T = 300$ K, $\mu_c = 0.8$ eV and $\hbar\gamma = 0.1$ meV. The gray vertical lines at $c\kappa/\omega = \sqrt{\varepsilon_1\mu_1}$ and $\sqrt{\varepsilon_2\mu_2}$ correspond to the modulus of the wave vector of light propagating above or below the graphene layer, whereas the gray vertical line at $c\kappa/\omega = 4$ corresponds to light propagating in a third medium, a prism, with $\varepsilon_p = 16$, $\mu_p = 1$.

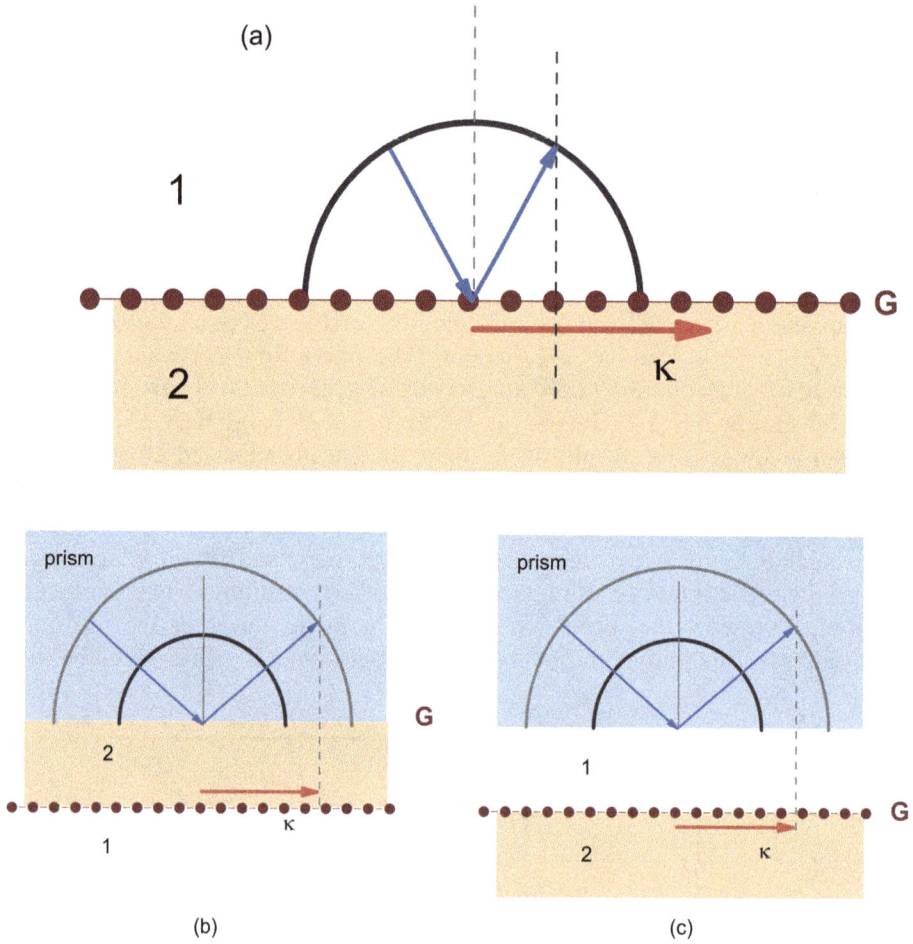

Figure 3.2. (a) The semicircle represents the dispersion relation for plane waves in medium 1. The projection along the interface of the wave vector of any incident plane wave is always greater than κ, the propagation constant for graphene SPPs. (b), (c) Using a third medium of higher refractive index $n_p = \sqrt{\varepsilon_p \mu_p} > c\kappa/\omega$, a coupling between the radiaton in the prism and the SPP propagating along the medium 1/graphene/medium 2 interface is expected.

of this semicircle, the projection along the interface of the wave vector of any incident plane wave is always smaller than κ, even at grazing incidence, where this projection takes its maximum value. Due to the conservation of the tangential component of the wave vector, the projection along the interface of the wave vector of the wave transmitted into medium 2 is also smaller than κ. For the Kubo parameters used in this case and for a frequency $\nu \approx 5.6$ THz (point A in figure 3.1), $\kappa \approx 3.25 \, \omega/c$, a value greater than the values that correspond to the wavenumbers of the plane waves in media 1 and 2. As a consequence of this mismatch between the tangential component of any wave vector of light propagating at any side of the graphene layer and the propagation constant of the graphene SPP, the periodic variation induced along the interface by any incident photon is always greater than the periodic variation required by the graphene surface plasmon, that is, phase-matching between photons and SPPs cannot occur at a smooth interface. Therefore, a photon from the dielectric above or below the graphene layer cannot excite an SPP, and neither can a graphene SPP emit energy as a photon into the surrounding dielectrics.

However, SPP–photon coupling may occur at a medium 1/graphene/medium 2 configuration if the projection along the interface of the incident wave vector is increased. One way to do this is illustrated in figure 3.2(b) and (c), where a third medium, usually a prism of refractive index $n_p = \sqrt{\varepsilon_p \mu_p}$, with $n_p > n_1$ and $n_p > n_2$, is introduced. If $n_p > c\kappa/\omega$ and the angle of incidence θ_0 between the incident wavevector and the normal to the graphene sheet is conveniently selected, a coupling between the radiaton in the prism and the SPP propagating along the medium 1/graphene/medium 2 interface is expected when

$$k_p \sin \theta_0 \approx \kappa, \tag{3.1}$$

that is, when the projection along the interface of the incident wave vector $k_p \sin \theta_0$ can be made equal to the propagation constant of the graphene SPP. This coupling condition is an approximated one, because in order to obtain the coupling, the simple configuration medium 1/graphene/medium 2 has necessarily been abandoned. It is clear that SPPs involved in the geometry illustrated in figure 3.2(b) and (c) are not strictly the same as those corresponding to the geometry illustrated in figure 3.2(a). For this reason, the correct interpretation of the coupling configurations requires us to consider layered media, that is, at least two interfaces ((b) and (c)) instead of only one (a).

The class of coupling geometries involving the introduction of higher index media to give the photon the additional momentum needed to achieve SPP excitation is known as *attenuated total internal reflection* (ATR) [1–3], a name coming from the fact that for $k_p \sin \theta_0$ to match κ, the fields in media 1 and 2 must be evanescent along the normal direction. Thus, the coupling mechanism in attenuated total reflection geometries involves tunneling of the incident field to the graphene sheet where the excitation of SPPs takes place. This tunneling is evidenced by the fact that the value of $c\kappa/\omega$ in the frequency–wavenumber diagram in figure 3.1 is to the right of values ck_1/ω and ck_2/ω. In order to excite the SPP at frequency $\nu \approx 5.6$ THz (point A in figure 3.1), a prism with $n_p > 3.25$ is required. Through using a prism with $\varepsilon_p = 16$, $\mu_p = 1$ (vertical line with abscissa $c\kappa/\omega = 4$), we expect to be able to excite SPPs at frequencies $\nu < 7.1$ THz. We observe that, if instead of point A in figure 3.1 we consider point B, corresponding to a frequency $\nu \approx 14.73$ THz and with a propagation constant $\kappa \approx 7.96 \, \omega/c$, the wavenumber mismatch between the graphene surface plasmons and plane waves in the media surrounding the graphene sheet is even larger than the mismatch at point A. Therefore, SPP excitation at the frequency corresponding to point B requires a prism with a higher index of refraction than the prism required for SPP excitation at the frequency corresponding to point B. As materials with a high value of n_p are not usually available, the larger the value of κ, the more difficult it is to excite graphene surface plasmons with attenuated total reflection. In these cases, it is preferable to use periodic structures, such as diffraction gratings, that change the tangential component of the incident wavevector by adding or subtracting an integer number of the grating wavevector (see chapter 4).

The control of the plasmonic response through an external gate voltage provided by ATR graphene-based structures makes them attractive in many applications, such as engineered absorption enhancement, graphene-based polarizers and optoelectronic switches [4–10].

3.2 Film coupler

The coupling condition equation (3.1) assumes that the characteristics of the surface plasmon polaritons excited along the graphene sheet in the presence of a prism are identical to those corresponding to a graphene sheet separating two semi-infinite regions. However, the correct interpretation of the excitation dynamics of graphene surface plasmon polaritons requires us to study the electromagnetic responses of structures such as those sketched in figure 3.2(b) and (c), that is, structures made up of at least three linear, isotropic and homogeneous media characterized by the constitutive parameters ε_i (electric permittivity) and μ_i (magnetic permeability), $i = 1, 2, 3$. As depicted in figure 3.3, we assume (a) a medium of incidence ($y > 0$, medium 3) with a positive refractive index $\nu_3 = \sqrt{\varepsilon_3 \mu_3}$ and (b) that a graphene monolayer located at $y = -d$ separates the region $-d < y < 0$ (medium 2) from the half-space $y < -d$ (medium 1, the medium of transmission).

As in section 2.1, we analyze two independent polarization cases separately: s (TE, electric field in the z direction) and p (TM, magnetic field in the z direction) and we denote by $f(x, y)$ the phasor of the z-directed component of the electromagnetic field. When this structure is illuminated by a linearly polarized,

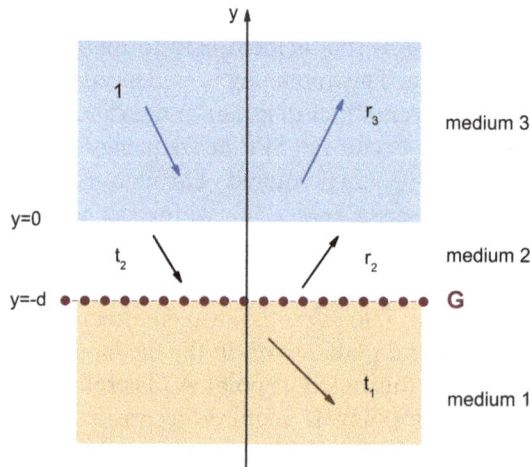

Figure 3.3. Attenuated total reflection scheme: the incident wave has unit amplitude, the graphene layer is located at $y = -d$ and the fields in each region are described by equation (3.2).

time-harmonic (angular frequency ω) plane wave propagating in the (x, y) plane and forming an angle θ_0 with the y axis, the phasor $f(x, y)$ in each region can be written as

$$
\begin{aligned}
f_3(x, y) &= e^{i(\kappa x - \beta_3 y)} + r_3 e^{i(\kappa x + \beta_3 y)} & y &> 0 \\
f_2(x, y) &= t_2 e^{i(\kappa x - \beta_2 y)} + r_2 e^{i(\kappa x + \beta_2 y)} & -d &< y < 0 \\
f_1(x, y) &= t_1 e^{i(\kappa x - \beta_1 y)} & y &< -d,
\end{aligned} \tag{3.2}
$$

where t_1, t_2, r_2 and r_3 are complex amplitudes, $\kappa = k_0 \nu_3 \sin \theta_0$, $k_0 = \omega/c$ is the projection along the interface of the wave vector of the incident plane wave and $\beta_j = \sqrt{\left(\frac{\omega}{c}\right)^2 \varepsilon_j \mu_j - \kappa^2}$ is the projection in the y-direction of the wave vector in each medium. Assuming that the transmission medium is a conventional positive refractive index medium, the condition $\operatorname{Im} \beta_j > 0$ must hold in each of the three regions.

3.2.1 *s* polarization

For *s* polarization, the boundary conditions (1.50) and (1.52) can be written in terms of $f(x, y)$ as

$$
\text{at } y = 0: \qquad f_3 = f_2, \qquad \frac{1}{\mu_2}\frac{\partial f_2}{\partial y} - \frac{1}{\mu_3}\frac{\partial f_3}{\partial y} = 0, \tag{3.3}
$$

$$
\text{at } y = -d: \qquad f_2 = f_1, \qquad \frac{1}{\mu_1}\frac{\partial f_1}{\partial y} - \frac{1}{\mu_2}\frac{\partial f_2}{\partial y} = i\omega\mu_0 \sigma f_1. \tag{3.4}
$$

Introducing expressions (3.2) into these boundary conditions, the following linear system of equations for the four unknown coefficients r_3, t_2, r_2 and t_1 is obtained:

$$
\begin{aligned}
r_3 + 1 &= t_2 + r_2, \\
t_2 e^{i\beta_2 d} + r_2 e^{-i\beta_2 d} &= t_1 e^{i\beta_1 d}, \\
\frac{\beta_3}{\mu_3}[r_3 - 1] &= \frac{\beta_2}{\mu_2}[r_2 - t_2], \\
-\frac{\beta_1}{\mu_1} t_1 e^{i\beta_1 d} - \frac{\beta_2}{\mu_2}[r_2 e^{-i\beta_2 d} - t_2 e^{i\beta_2 d}] &= \omega\mu_0 \sigma t_1 e^{i\beta_1 d}.
\end{aligned} \tag{3.5}
$$

In matrix notation, system equation (3.5) is rewritten as

$$
\begin{bmatrix}
-1 & 1 & 1 & 0 \\
0 & e^{i\beta_2 d} & e^{-i\beta_2 d} & -e^{i\beta_1 d} \\
\dfrac{\beta_3}{\mu_3} & \dfrac{\beta_2}{\mu_2} & -\dfrac{\beta_2}{\mu_2} & 0 \\
0 & -\dfrac{\beta_2}{\mu_2}e^{i\beta_2 d} & \dfrac{\beta_2}{\mu_2}e^{-i\beta_2 d} & \left(\dfrac{\beta_1}{\mu_1} + \omega\mu_0\sigma\right)e^{i\beta_1 d}
\end{bmatrix}
\begin{bmatrix} r_3 \\ t_2 \\ r_2 \\ t_1 \end{bmatrix}
=
\begin{bmatrix} 1 \\ 0 \\ \dfrac{\beta_3}{\mu_3} \\ 0 \end{bmatrix}. \tag{3.6}
$$

The second and fourth rows of system equation (3.6) represent the boundary conditions at a graphene layer located at $y = -d$ and separating media 1 and 2. These are the same boundary conditions that were considered in section 2.1.1 for the Fresnel problem at a graphene layer located at $y = 0$ and separating media 1 and 2, except that the role of the incident, reflected and transmitted waves in section 2.1.1 is played here by the waves with the amplitudes t_2, r_2 and t_1, respectively. Therefore, except for a phase shift due to the change of origin, r_2 and t_1 have essentially the same form as r^s in equation (2.8) and t^s in equation (2.9):

$$r_2 = \frac{\beta_2/\mu_2 - \beta_1/\mu_1 - \omega\mu_0\sigma}{\beta_2/\mu_2 + \beta_1/\mu_1 + \omega\mu_0\sigma}e^{i2\beta_2 d}t_2 \tag{3.7}$$

$$t_1 = \frac{2\beta_2/\mu_2}{\beta_2/\mu_2 + \beta_1/\mu_1 + \omega\mu_0\sigma}e^{i(\beta_2-\beta_1)d}t_2. \tag{3.8}$$

In order to make the connection with the Fresnel problem for two half-spaces more explicit, we use the alternative notation introduced in section 2.1.3. With this notation, r_{igj} and t_{igj} are the amplitude reflection and transmission coefficients for waves impinging from the i side into a single graphene layer separating medium i from medium j, whereas r_{ij} and t_{ij} are the coefficients corresponding to a bare (that is, without a graphene layer) interface between medium i and medium j. Noting that the fraction in equation (3.7) coincides with r_{2g1}, and that the fraction in equation (3.8) coincides with t_{2g1}, equations (3.7) and (3.8) can be rewritten as

$$r_2 = r_{2g1}e^{i2\beta_2 d}t_2 \qquad \text{and} \qquad t_1 = t_{2g1}e^{i(\beta_2-\beta_1)d}t_2. \tag{3.9}$$

Using these results and the boundary conditions at $y = 0$ represented by the first and third rows of system equation (3.6), the following expressions for the unknown amplitudes of the electric field represented by equation (3.2) are finally obtained:

$$r_3 = \frac{r_{32} + r_{2g1}e^{i2\beta_2 d}}{1 + r_{32}r_{2g1}e^{i2\beta_2 d}} \tag{3.10}$$

$$t_2 = \frac{t_{32}}{1 + r_{32}r_{2g1}e^{i2\beta_2 d}} \tag{3.11}$$

$$r_2 = \frac{t_{32}r_{2g1}e^{i2\beta_2 d}}{1 + r_{32}r_{2g1}e^{i2\beta_2 d}} \tag{3.12}$$

$$t_1 = \frac{t_{32}t_{2g1}e^{i(\beta_2-\beta_1)d}}{1 + r_{32}r_{2g1}e^{i2\beta_2 d}}. \tag{3.13}$$

These expressions for r_3 and t_1 in terms of the coefficients r_{2g1} and t_{2g1} for a single graphene layer and the coefficients r_{32} and t_{32} of a graphene-free interface have the same form as the usual expressions for the amplitude reflection and transmission coefficients of a semi-transparent plate without infinitesimally thin current sheets at its boundaries (see equations (2.30) and (2.31)). This is because, despite the different boundary conditions satisfied with or without graphene, the procedure used in optics in the frame of multiple beam interference at semi-transparent plates is valid independently of the presence of a graphene layer at one of the plate boundaries. The standard procedure [11] assumes that the incident wave is divided into two plane waves at the first interface, one reflected into the medium of incidence and the other transmitted into the plate. The latter wave is incident on the second surface—in our case the graphene layer separating media 2 and 1—and then divided into two plane waves, one transmitted into the region below the plate (region 1) and the other reflected back into the plate. Taking into account the fact that the process for the division of the waves that remain inside the plate continues *ad infinitum*, the expressions for amplitudes r_3, t_2, r_2 and t_1 can be expressed as geometrical series which only assume knowledge of the reflection and transmission coefficients for a single isotropic interface between two isotropic media, but do not depend on the details of the physical mechanisms producing reflection and transmission at each plate boundary.

3.2.2 *p* polarization

For *p*–polarized incident waves, the boundary conditions (1.50) and (1.52) are

$$\frac{1}{\varepsilon_3}\frac{\partial f_3}{\partial y} = \frac{1}{\varepsilon_2}\frac{\partial f_2}{\partial y}, \qquad f_2 - f_3 = 0, \tag{3.14}$$

at $y = 0$, and

$$\frac{1}{\varepsilon_2}\frac{\partial f_2}{\partial y} = \frac{1}{\varepsilon_1}\frac{\partial f_1}{\partial y}, \qquad f_1 - f_2 = -i\frac{\sigma}{\omega\varepsilon_0}\frac{1}{\varepsilon_1}\frac{\partial f_1}{\partial y}, \tag{3.15}$$

at $y = -d$. These boundary conditions lead to the following linear system for the unknown amplitudes r_1, t_2, r_2 and t_3

$$r_3 + 1 = t_2 + r_2,$$

$$t_1 e^{i\beta_1 d} - t_2 e^{i\beta_2 d} - r_2 e^{-i\beta_2 d} = -\frac{\sigma}{\omega\varepsilon_0}\frac{\beta_1}{\varepsilon_1}t_1 e^{i\beta_1 d},$$

$$\frac{\beta_3}{\varepsilon_3}[r_3 - 1] = \frac{\beta_2}{\varepsilon_2}[r_2 - t_2], \tag{3.16}$$

$$\frac{\beta_2}{\varepsilon_2}[r_2 e^{-i\beta_2 d} - t_2\, e^{i\beta_2 d}] = -\frac{\beta_1}{\varepsilon_1}t_1 e^{i\beta_1 d}.$$

In matrix notation, system (3.16) can be rewritten as

$$
\begin{pmatrix}
-1 & 1 & 1 & 0 \\[2mm]
0 & e^{i\beta_2 d} & e^{-i\beta_2 d} & -\left(1 + \dfrac{\sigma}{\omega}\dfrac{\beta_1}{\varepsilon_0 \varepsilon_1}\right)e^{i\beta_1 d} \\[4mm]
\dfrac{\beta_3}{\varepsilon_3} & \dfrac{\beta_2}{\varepsilon_2} & -\dfrac{\beta_2}{\varepsilon_2} & 0 \\[4mm]
0 & -\dfrac{\beta_2}{\varepsilon_2}e^{i\beta_2 d} & \dfrac{\beta_2}{\varepsilon_2}e^{-i\beta_2 d} & \dfrac{\beta_1}{\varepsilon_1}e^{i\beta_1 d}
\end{pmatrix}
\begin{pmatrix} r_3 \\ t_2 \\ r_2 \\ t_1 \end{pmatrix}
=
\begin{pmatrix} 1 \\ 0 \\ \dfrac{\beta_3}{\varepsilon_3} \\ 0 \end{pmatrix}.
\tag{3.17}
$$

As in the s polarization case, the second and fourth rows of this system represent the boundary conditions at a graphene layer located at $y = -d$ separating media 2 and 1 and correspond to the Fresnel problem considered for a graphene layer in section 2.1.2. In this correspondence, the waves with amplitudes t_2, r_2 and t_1 play exactly the same role as the incident, reflected and transmitted waves in section 2.1.2, respectively. Therefore, r_2 and t_1 result in

$$
r_2 = \frac{\varepsilon_1/\beta_1 - \varepsilon_2/\beta_2 + \sigma/\omega\varepsilon_0}{\varepsilon_2/\beta_2 + \varepsilon_2/\beta_2 + \sigma/\omega\varepsilon_0}e^{i2\beta_2 d}t_2
\tag{3.18}
$$

$$
t_1 = \frac{2\varepsilon_1/\beta_1}{\varepsilon_1/\beta_1 + \varepsilon_2/\beta_2 + \sigma/\omega\varepsilon_0}e^{i(\beta_2 - \beta_1)d}t_2,
\tag{3.19}
$$

with essentially the same form as r^p in equation (2.13) and t^p in equation (2.14), except for the phase shift $e^{i2\beta_2 d}$, resulting from the change of origin, and a subscript interchange. Using the p polarization Fresnel coefficients for a graphene sheet separating two half-spaces, equations (3.18) and (3.19) become

$$
r_2 = r_{2g1}e^{i2\beta_2 d}t_2 \qquad \text{and} \qquad t_1 = t_{2g1}e^{i(\beta_2 - \beta_1)d}t_2.
\tag{3.20}
$$

Using these results and the boundary conditions at $y = 0$ represented by the first and third rows of system equation (3.17), we obtain that the four unknown amplitudes of the magnetic field represented by equation (3.2) are given formally by the same expressions obtained in equations (3.10)–(3.13) for s polarization. The formal identity between the amplitudes of $f(x, y)$ in both polarization cases, and the formal identity between the amplitudes of $f(x, y)$ with or without graphene at the plate boundaries discussed at the end of section 3.2.2, results from the validity of the Airy summation process and does not depend on the details of the electromagnetic mechanisms producing reflection and transmission at each plate boundary, although the reflection and transmission coefficients for a single isotropic interface do depend on the details of the electromagnetic interactions at the interface.

3.3 ATR excitation of graphene SPPs

3.3.1 s polarization

According to the discussion in section 2.3.2, the condition Im $\tilde{\sigma} < 0$ should hold at the frequency of interest for s-polarized SPPs to exist on a conductivity sheet.

However, it has been shown [12] that even when this condition holds, efficient coupling between an incident wave and s-polarized SPPs presents many experimental challenges. One difficulty is related to the existence of a critical value for the film thickness d between the coupling prism and the graphene layer. Below this value, the s-polarized SPPs in the coupling configuration sketched in figure 3.2(a) and (b) are no longer fully bound to the graphene sheet. This critical value can be obtained from the dispersion relation

$$e^{i2\beta_2 d} = -1/(r_{32}\, r_{2g1})$$

resulting from requiring the propagation constant κ to be a complex pole of the plane wave coefficients in equations (3.10)–(3.13). Using the expressions for the amplitude reflection coefficients r_{32} and r_{2g1} corresponding to a single interface, the dispersion relation for s-polarized SPPs in the ATR configurations becomes

$$e^{i2\beta_2 d} = -\frac{\beta_3/\mu_3 + \beta_2/\mu_2}{\beta_3/\mu_3 - \beta_2/\mu_2}\frac{\beta_2/\mu_2 + \beta_1/\mu_1 + 4\pi\omega\alpha\tilde{\sigma}/c}{\beta_2/\mu_2 - \beta_1/\mu_1 - 4\pi\omega\alpha\tilde{\sigma}/c}. \qquad (3.21)$$

This equation can be used to show [12] that the cutoff value for the thickness d between the coupling prism and the graphene sheet is inversely proportional to $|\operatorname{Im}\tilde{\sigma}|$. Figure 1.1 shows that the most negative value of $\operatorname{Im}\tilde{\sigma}$ is taken at $T = 0$ K and that typical values at room temperature have $|\operatorname{Im}\tilde{\sigma}| \lesssim 0.25$. Therefore, large cutoff thickness results, which prevents efficient resonant coupling between the incident radiation and s-polarized SPPs in this ATR configuration.

The theoretical simulations for realistic values of $\tilde{\sigma}$ given in [12] illustrate the experimental challenges—such as the very low contrast and exceptionally narrow angular width of the ATR minimum—imposed by the unusual features of s-polarized graphene SPPs. Here we choose $\tilde{\sigma}$ with an imaginary part that is slightly greater than the realistic values predicted by the Kubo model. Although these values do not correspond to actual graphene single sheets at room temperature, they can be used to illustrate the ATR detection mechanism for s-polarized graphene SPPs and can be thought of as a representation of the effective increase of the conductivity observed in graphene multilayers [13].

Let us assume that $\tilde{\sigma} = 0.125 - i$ and that the conductivity sheet is immersed in a medium with $\varepsilon = 2.25$ and $\mu = 1$. From equation (2.41) we obtain $\kappa \approx (1.501 + 0.0002i)\omega/c$, that is, a propagation constant with a real part that is slightly greater than the value $\omega\sqrt{\varepsilon\mu}/c$. Taking into account the fact that $\pi^2\alpha^2 \approx 0.02$, equation (2.49) shows that the value of κ does not change appreciably when the value of $\operatorname{Im}\tilde{\sigma}$ is doubled to -2. In order to obtain coupling between the incident radiation and s-polarized SPPs at a conductivity sheet with such values of $\tilde{\sigma}$, we introduce a prism with $\varepsilon_p = 4$ and $\mu_p = 1$, as sketched in figure 3.2. According to the approximated condition equation (3.1), for the value of the parameters used in this example the coupling is expected for $\sin\theta_0 \approx \kappa/k_p$, that is, for angles of incidence $\theta_0 \approx 48.61°$. As this condition cannot give information on the dynamics of the SPP–photon coupling, we resort to the results obtained in section 3.2.1. Figures 3.4 (for $\tilde{\sigma} = 0.125 - i$) and 3.5 (for $\tilde{\sigma} = 0.125 - 2i$) show curves of reflectivity $R = |r_3|^2$

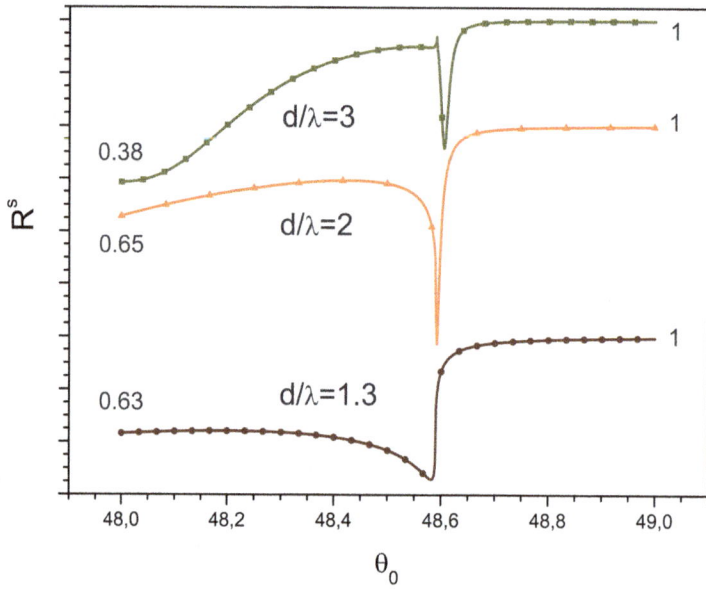

Figure 3.4. Stacked curves of s polarization reflectance $R = |r_3|^2$ (see equation (3.10)) as functions of the angle of incidence for an ATR configuration with $\varepsilon_1 = \varepsilon_2 = 2.25$, $\varepsilon_3 = 4$, three different values of the film thickness $d/\lambda = 1.3$, 2 and 3, and $\tilde{\sigma} = 0.125 - i$. The reflectance values corresponding to the beginning and end points indicate the scale for each curve.

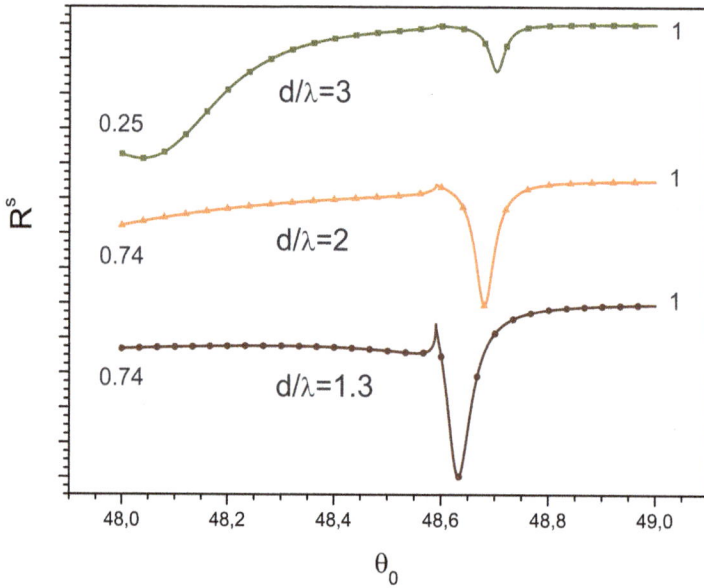

Figure 3.5. As in figure 3.4, but for $\tilde{\sigma} = 0.125 - 2i$.

versus θ_0 obtained from equation (3.10) with $\varepsilon_1 = \varepsilon_2 = 2.25$, $\varepsilon_3 = 4$, $\mu_1 = \mu_2 = \mu_3 = 1$ for angles of incidence near the value predicted by the approximated condition equation (3.1) and for three different values ($d/\lambda = 1.3, 2, 3$) for the thickness of the film between the coupling prism and the graphene layer. The Python script used to calculate these reflectivities curves is given in the appendix.

The curves in figures 3.4 and 3.5 exhibit a minimum at a position that is very close to but not equal to the estimate given by equation (3.1). Although the existence of such a reflectivity minimum in phase-matching configurations is a manifestation of SPP excitation, it has been noted [14, 15] that in these configurations maximal absorption of incident waves and maximal intensity of the excited SPPs do not always occur under the same conditions, a feature that indicates that caution should be exercised when we obtain maps of SPP dispersion relations by optically detecting the angular position of the minima in prism and grating couplers. Theoretically, the correct dispersion relation can be obtained by finding the complex roots of equation (3.21). Analogously, the angular position for the maximal absorption of incident waves can be obtained by finding the complex zeroes of equation (3.10). It is the position of these two complex values, the complex pole and the complex zero, that ultimately determines the symmetry (as in the curves for $d/\lambda = 2$ and 3 in figure 3.4) or asymmetry (as in the curve for $d/\lambda = 1.3$ in figure 3.4) of the reflectance curves around the position of the resonant coupling [14, 16, 17]. Details of the pole-zero treatment, which can be applied to s- and p-polarized SPPs, are omitted here for reasons of space, and the reader is referred to [14, 16, 17] for further information.

3.3.2 p polarization

The values in the frequency–wavenumber diagram for p-polarized SPPs given in figure 3.1 were obtained by neglecting the small real part of $\tilde{\sigma}$. However, for the frequency corresponding to point A ($\nu \approx 5.6$ THz) and for Kubo parameters $T = 300$ K, $\mu_c = 0.8$ eV and $\hbar\gamma = 0.1$ meV, the surface conductivity takes the value $\tilde{\sigma} \approx 0.05 + 10.83\,i$, and therefore the correct value for the propagation constant at this frequency has a small imaginary part and is $\kappa \approx (3.29 + 0.01\,i)\,\omega/c$.

Assuming that in order to match this value of κ we use a prism with refractive index $n_p = 4$ ($\varepsilon_p = 16$, $\mu_p = 1$), the approximate condition equation (3.1) predicts a coupling at $\theta_0 \approx 55.34°$, with the critical angle of total reflection being $\theta_0 \approx 14.48°$. To investigate the p polarization coupling at the graphene layer, we use the theoretical results obtained in section 3.2.

In figure 3.6 we show stacked curves of p polarization reflectance $R = |r_3|^2$ versus θ_0 calculated with equation (3.10) using $\varepsilon_1 = 1$, $\varepsilon_2 = 2$, $\varepsilon_3 = 16$, $\mu_1 = \mu_2 = \mu_3 = 1$ for angles of incidence greater than the critical angle of total reflection and for five different values ($d/\lambda = 0.05, 0.075, 0.1, 0.15,$ and 0.2) for the thickness of the film between the coupling prism and the graphene layer. The Python script used to obtain these curves can be found in the appendix. As in the s polarization examples considered in figures 3.4 and 3.5, all the reflectance curves in figure 3.6 exhibit a minimum in an angular range where total reflection is expected for the same structure without a graphene sheet. The positions of these minima depend on the

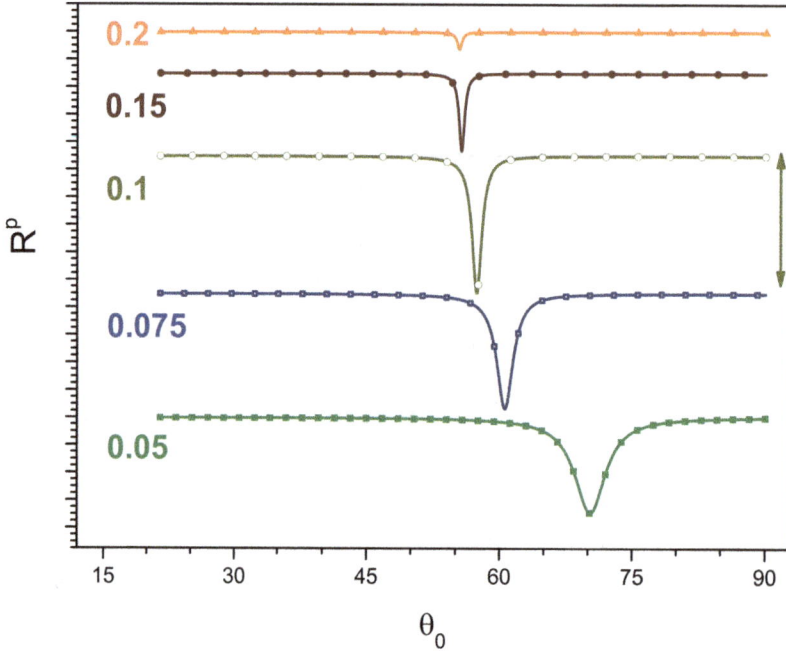

Figure 3.6. Stacked curves of p polarization reflectance $R = |r_3|^2$ (see equation (3.10)) as functions of the angle of incidence for ATR configurations with $\varepsilon_1 = 1$, $\varepsilon_2 = 2$, $\varepsilon_3 = 16$ and five different values of film thickness ($d/\lambda = 0.05$, 0.075, 0.1, 0.15, and 0.2). The graphene layer is characterized by $\tilde{\sigma} = 0.0486529 + 10.8258i$, a value obtained from the Kubo model for $T = 300$ K, $\mu_c = 0.8$ eV and $\hbar\gamma = 0.1$ meV. The gray vertical double arrow close to the curve for $d/\lambda = 0.1$ corresponds to unity reflectance variation and indicates the reflectance scale. The Python script used to obtain these curves can be found in the appendix.

film thickness and are $\approx 70.24°$, $60.57°$, $57.39°$, $55.65°$ and $55.35°$ for $d/\lambda = 0.05$, 0.075, 0.1, 0.15 and 0.2, respectively. We observe that, in contrast to the s polarization examples, the angular positions of the minima are not always very close to the value predicted by the approximated condition (3.1) ($\theta_0 \approx 55.34°$ in this case), particularly for the lower values of d/λ, where the propagation characteristics of the dielectric/graphene/dielectric interface are more strongly affected by the presence of the prism. For the curves corresponding to $d/\lambda = 0.15$ and 0.2, on the other hand, the close coincidence between the predicted and observed angular positions of the minima indicates that the propagation characteristics of the dielectric/graphene/dielectric interface are much less affected by the presence of the prism, although the coupling is much less efficient. Near-maximal absorption of the incident wave occurs for $d/\lambda = 0.1$, where the SPP–photon coupling produces near-unity reflectance variation. As noted in the discussion of the results illustrating s polarization coupling, maximal absorption of incident waves and maximal intensity of the excited SPPs do not always occur under the same conditions in phase-matching techniques [14, 15], and these generally different conditions can be obtained by finding the complex zeroes and poles of the reflection coefficient [14, 16, 17].

3.4 Periodic multilayers

Periodic arrangements of two or more different materials are called photonic crystals [18, 19] or photonic band gap (PBG) structures [20, 21]. These structures exhibit frequency selective response properties: there are frequency ranges (called photonic band gaps or stopbands) in which propagation is forbidden, and there are frequency ranges (called passbands) in which transmission is allowed.

A periodic stack of layers or sheets, called a periodic multilayer, is the simplest PBG structure. The characteristics of the photonic band gap structure of a periodic multilayer are highly dependent on the type of materials used in its fabrication. Therefore, the possible incorporation of new photonic materials in PBG structures has always been considered an attractive idea. Coinciding with the rise of optical metamaterials, the study of electromagnetic propagation through periodic multilayers containing magnetic metamaterials has revealed the existence of two types of new band gaps that are not present in multilayers of conventional nonmagnetic optical materials [22], which are invariant to scale-change and robust against disorder: the zero-index band gap [23, 24] and the μ-zero and ε-zero band gaps [25–27]. Periodic multilayers in which graphene sheets are embedded between adjacent dielectric layers have also attracted the attention of researchers and recent studies [28–31] have shown that, apart from the usual structural Bragg gaps, graphene–dielectric multilayers exhibit a graphene-induced photonic band gap, while both kinds of gaps can be tuned by a gate voltage. In addition to their spectral tunability, multi-layer graphene stacks could provide a platform to improve the low detectability of s-polarized SPPs [12, 13, 32–34], and this is also of interest to researchers. A general procedure for finding the electromagnetic normal modes (a homogeneous problem, with no incident waves) in a different kind of layered structure—including graphene and graphene-like sheets—was recently presented in [35]. We finish this chapter by reviewing the use of the transfer matrix method [22] to study the reflection and transmission of a wave at periodic planar multilayers containing graphene sheets.

3.4.1 N layers: transfer matrix

The film analyzed in section 3.2 corresponds to the case of a single homogeneous layer (the film) surrounded by two homogeneous half-spaces, with a graphene sheet located at only one of the two film interfaces. In what follows we first consider the more general case in which a film of a homogeneous material of type B is bounded by two identical graphene sheets [28]. Second, we construct a locally periodic structure consisting of N parallel layers of type B periodically distributed in a homogeneous medium of type A. A sketch of this locally periodic structure is shown in figure 3.7. Layers of type B have thicknesses d_B and contain an isotropic dielectric material with constitutive parameters ε_B, μ_B, while the graphene sheets are characterized by a surface conductivity $\sigma(\omega)$ with identical Kubo parameters. The layers are parallel to the (xz) plane and the y axis is the stratification direction. The graphene-bounded layers are immersed in medium A, with constitutive parameters ε_A, μ_A and the structure has period $d = d_A + d_B$.

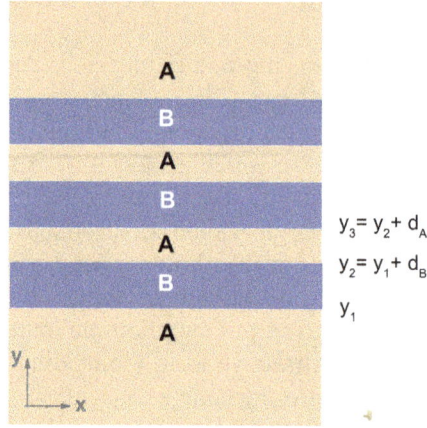

Figure 3.7. The case $N = 3$ for periodic structures of type $(AGBG)^N$, consisting of N parallel layers of a material of type B periodically distributed in a homogeneous medium of type A and with identical graphene sheets embedded between consecutive layers.

If the incident wave arrives from the region below the bottom layer, the reflected wave propagates in the same region towards $y \to -\infty$ (downward in figure 3.7) and there is no wave moving towards $y \to -\infty$ in the medium of transmission. Similarly, if the incident wave arrives from the region above the top layer, the reflected wave propagates in the same region towards $y \to +\infty$ (upward in figure 3.7) and there is no wave moving towards $y \to +\infty$ in the region below the bottom layer. In every other region different from the region of incidence and transmission, there are two waves, one propagating in the direction of the incident wave and another in the direction of the reflected wave. We assume that this structure is illuminated by a linearly polarized time-harmonic (angular frequency ω) plane wave propagating in the (x, y) plane and forming an angle θ_0 with the y axis.

Let us consider the unit cell configuration AGBGA, that is, a configuration consisting of the following consecutive elements: a region with medium A, a graphene sheet, a region with medium B, another identical graphene sheet and a region with medium A. The phasor $f(x, y)$ denoting the z-directed component of the electromagnetic field in three consecutive regions bounded by graphene sheets at y_1 and $y_2 = y_1 + d_B$ can be written as

$$
\begin{aligned}
f(x, y) &= (A_1 e^{-i\beta_A y} + B_1 e^{i\beta_A y}) \, e^{i\kappa x} & y &< y_1 \\
f(x, y) &= (A_2 e^{-i\beta_B y} + B_2 e^{i\beta_B y}) \, e^{i\kappa x} & y_1 &< y < y_2, \\
f(x, y) &= (A_3 e^{-i\beta_A y} + B_3 e^{i\beta_A y}) \, e^{i\kappa x} & y_2 &< y
\end{aligned}
\tag{3.22}
$$

where $\beta_A^2 = \omega^2 \epsilon_A \mu_A / c^2 - \kappa^2$, $\beta_B^2 = \omega^2 \epsilon_B \mu_B / c^2 - \kappa^2$ and $d = d_A + d_B$, the period of the structure.

The boundary conditions at the graphene sheet located at $y = y_1$ have the same form as equation (3.4) in the s polarization case, or equation (3.15) in the p polarization case. These boundary conditions can be used to provide relations between amplitudes A_1 and B_1 in the first layer and amplitudes A_2 and B_2 in the second layer. For example, after a slight change in notation, equation (3.4) for s polarization give

$$A_1 e^{-i\beta_A y_1} + B_1 e^{i\beta_A y_1} = A_2 e^{-i\beta_B y_1} + B_2 e^{i\beta_B y_1}$$

$$-A_1 \frac{\beta_A}{\mu_A} e^{-i\beta_A y_1} + B_1 \frac{\beta_A}{\mu_A} e^{i\beta_A y_1} = -A_2 \left(\frac{\beta_B}{\mu_B} - \omega\mu_0\sigma \right) e^{-i\beta_B y_1} + B_2 \left(\frac{\beta_B}{\mu_B} + \omega\mu_0\sigma \right) e^{i\beta_B y_1},$$

or, in matrix form

$$\underbrace{\begin{bmatrix} e^{-i\beta_A y_1} & e^{i\beta_A y_1} \\ -\frac{\beta_A}{\mu_A} e^{-i\beta_A y_1} & \frac{\beta_A}{\mu_A} e^{i\beta_A y_1} \end{bmatrix}}_{\mathbf{M}_{A1}} \begin{bmatrix} A_1 \\ B_1 \end{bmatrix} = \underbrace{\begin{bmatrix} e^{-i\beta_B y_1} & e^{i\beta_B y_1} \\ -\left(\frac{\beta_B}{\mu_B} - \omega\mu_0\sigma \right) e^{-i\beta_B y_1} & \left(\frac{\beta_B}{\mu_B} + \omega\mu_0\sigma \right) e^{i\beta_B y_1} \end{bmatrix}}_{\mathbf{M}_{B1}} \begin{bmatrix} A_2 \\ B_2 \end{bmatrix}. \qquad (3.23)$$

The boundary conditions at the graphene sheet located at $y = y_2$ provide analogous matrix relations between amplitudes A_2 and B_2 in the second layer and amplitudes A_3 and B_3 in the third layer:

$$\underbrace{\begin{bmatrix} e^{-i\beta_B y_2} & e^{i\beta_B y_2} \\ -\frac{\beta_B}{\mu_B} e^{-i\beta_B y_2} & \frac{\beta_B}{\mu_B} e^{i\beta_B y_2} \end{bmatrix}}_{\mathbf{M}_{B2}} \begin{bmatrix} A_2 \\ B_2 \end{bmatrix} = \underbrace{\begin{bmatrix} e^{-i\beta_A y_2} & e^{i\beta_A y_2} \\ -\left(\frac{\beta_A}{\mu_A} - \omega\mu_0\sigma \right) e^{-i\beta_A y_2} & \left(\frac{\beta_A}{\mu_A} + \omega\mu_0\sigma \right) e^{i\beta_A y_2} \end{bmatrix}}_{\mathbf{M}_{A2}} \begin{bmatrix} A_3 \\ B_3 \end{bmatrix}. \qquad (3.24)$$

Thus, using equations (3.23) and (3.33), we can express the column vector with amplitudes A_3 and B_3 of the waves in the third layer in terms of the column vector with amplitudes A_1 and B_1 of the waves in the first layer as

$$\begin{bmatrix} A_3 \\ B_3 \end{bmatrix} = \mathbf{M} \begin{bmatrix} A_1 \\ B_1 \end{bmatrix}, \qquad (3.25)$$

where

$$\mathbf{M} = [\mathbf{M}_{A2}]^{-1} \mathbf{M}_{B2} [\mathbf{M}_{B1}]^{-1} \mathbf{M}_{A1} \qquad (3.26)$$

is the *transfer matrix* that characterizes the unit cell of the AGBGA locally periodic structure sketched in figure 3.7. A similar relation holds for p polarization, but now the matrices involved in the boundary conditions are

$$\mathbf{M}_{A1} = \begin{bmatrix} e^{-i\beta_A y_1} & e^{i\beta_A y_1} \\ -\frac{\beta_A}{\varepsilon_A} e^{-i\beta_A y_1} & \frac{\beta_A}{\varepsilon_A} e^{i\beta_A y_1} \end{bmatrix} \qquad (3.27)$$

$$\mathbf{M}_{B1} = \begin{bmatrix} \left(1 - \dfrac{\beta_B \sigma}{\omega \varepsilon_0 \varepsilon_B}\right) e^{-i\beta_B y_1} & \left(1 + \dfrac{\beta_B \sigma}{\omega \varepsilon_0 \varepsilon_B}\right) e^{i\beta_B y_1} \\[2ex] -\dfrac{\beta_B}{\varepsilon_B} e^{-i\beta_A y_1} & \dfrac{\beta_B}{\varepsilon_B} e^{i\beta_B y_1} \end{bmatrix} \tag{3.28}$$

$$\mathbf{M}_{A2} = \begin{bmatrix} \left(1 - \dfrac{\beta_A \sigma}{\omega \varepsilon_0 \varepsilon_A}\right) e^{-i\beta_A y_2} & \left(1 + \dfrac{\beta_A \sigma}{\omega \varepsilon_0 \varepsilon_A}\right) e^{i\beta_A y_2} \\[2ex] -\dfrac{\beta_A}{\varepsilon_A} e^{-i\beta_A y_2} & \dfrac{\beta_A}{\varepsilon_A} e^{i\beta_A y_2} \end{bmatrix} \tag{3.29}$$

$$\mathbf{M}_{B2} = \begin{bmatrix} e^{-i\beta_B y_2} & e^{i\beta_B y_2} \\[2ex] -\dfrac{\beta_B}{\varepsilon_B} e^{-i\beta_B y_2} & \dfrac{\beta_B}{\varepsilon_B} e^{i\beta_B y_2} \end{bmatrix}. \tag{3.30}$$

For normal incidence, $\kappa = 0$, the transfer matrix for s polarization and the transfer matrix for p polarization are identical.

The expressions above allow us to obtain the AGBGA transfer matrix that gives the column vector for the phasor amplitudes in a medium at one side of an inhomogeneity in the form of a graphene-sandwiched dielectric in terms of the column vector for the phasor amplitudes *in the same medium* at the other side of the inhomogeneity. Invoking the appropriate boundary conditions, similar transfer matrices can be constructed for more complicated inhomogeneities of the type $A(G_1BG_2CG_3 \ldots)A$.

Because of geometric and electromagnetic symmetries, and regardless of the specific form of the inhomogeneity, all transfer matrices share common properties, such as the fact that their determinant is always equal to one. In addition, when the inhomogeneity is symmetric, as in the simple case of the AGBGA configuration considered here, the diagonal elements of the transfer matrix satisfy the relation $M_{12} = -M_{21}$. Proofs of these properties in the context of quantum mechanical scattering in one dimension can be seen in [36]. An alternative proof can be given by applying relation equation (3.25) to the two configurations sketched in figure 3.8. In one configuration, the incident wave arrives from below the inhomogeneity with unit amplitude $B_1 = 1$, and there is no incident wave in the region above the inhomogeneity, $A_3 = 0$. Thus, $A_1 = r$ and $B_3 = t$, the plane wave reflection and transmission coefficient of the inhomogeneity, respectively. Enforcing equation (3.25) with these amplitudes, we obtain

$$\left. \begin{aligned} 0 &= M_{11}r + M_{12} \\ t &= M_{21}r + M_{22} \end{aligned} \right\}. \tag{3.31}$$

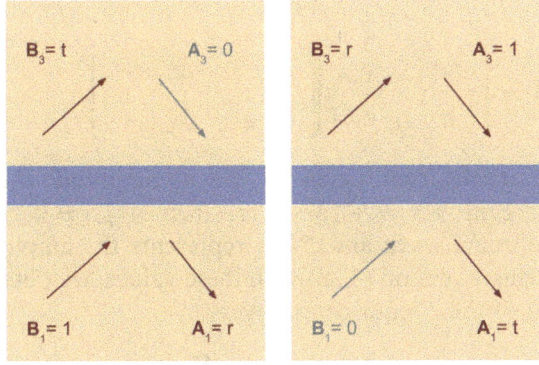

Figure 3.8. The reflected and transmitted plane wave coefficients for a symmetric inhomogeneity immersed in a homogeneous medium. In (a) the incident wave arrives from below the inhomogeneity with unit amplitude and there is no incident wave in the region above the inhomogeneity, whereas in (b) the incident wave with unit amplitude arrives from above and there is no incident wave below the inhomogeneity.

Analogously, in the second configuration the incident wave arrives from above the inhomogeneity with unit amplitude $A_3 = 1$, and there is no incident wave in the region below the inhomogeneity, $B_1 = 0$. Thus, as both configurations are indistinguishable, $A_1 = t$ and $B_3 = r$, the same plane wave coefficients correspond to the configuration sketched in figure 3.8(b). Enforcing equation (3.25) with these new amplitude column vectors, we obtain

$$\left. \begin{array}{l} 1 = M_{11}t \\ r = M_{21}t \end{array} \right\}. \tag{3.32}$$

The same arguments are valid for all symmetric inhomogeneities. The proof that in this case the determinant of the transfer matrix is equal to one and $M_{12} = -M_{21}$ concludes when the plane wave coefficients r and t are eliminated from equations (3.31) and (3.32).

3.4.2 N layers: reflectivity and transmissivity

The reflectivity and transmissivity of a structure consisting of N equally spaced graphene-bounded layers of medium B inmersed in medium A, the $(AB)^N$ structure sketched in figure 3.7, and the band structure of an ideal perfectly periodic distribution with an infinite number of such layers, can be obtained from the transfer matrix $\mathbf{M}$. The fields $f(x, y)$ in the bottom and top half-spaces have amplitudes A_1 and B_1 in region $y < y_1$, and A_{2N+1} and B_{2N+1} in region $y > y_2 + (N - 1)d$ From the definition of $\mathbf{M}$ in equation (3.26), it follows that the column vector with amplitudes A_{2N+1} and B_{2N+1} is related to the column vector with amplitudes A_1 and B_1 through a matrix $\mathbf{T}$ [36]

$$\begin{bmatrix} A_{2N+1} \\ B_{2N+1} \end{bmatrix} = \begin{bmatrix} T_{11} & T_{12} \\ T_{21} & T_{22} \end{bmatrix} \begin{bmatrix} A_1 \\ B_1 \end{bmatrix}, \tag{3.33}$$

with

$$\mathbf{T} = \begin{bmatrix} e^{i\beta_A Nd} & 0 \\ 0 & e^{-i\beta_A Nd} \end{bmatrix} \begin{bmatrix} M_{11}e^{-i\beta_A d} & M_{12}e^{-i\beta_A d} \\ M_{21}e^{i\beta_A d} & M_{22}e^{i\beta_A d} \end{bmatrix}^N. \tag{3.34}$$

If the incident wave arrives from region $y < y_1$ with amplitude B_1, and if there is no incident wave in region $y > y_2 + (N-1)d$, then $A_{2N+1} = 0$, A_1 represents the amplitude of the reflected wave and B_{2N+1} represents the amplitude of the transmitted wave. Enforcing equation (3.33) with these values, we obtain the plane wave reflection coefficient of the N equispaced layers

$$r_N = \left.\frac{A_1}{B_1}\right)_{A_{2N+1}=0} = -\frac{T_{12}}{T_{11}}. \tag{3.35}$$

Similarly, if the incident wave arrives from region $y > y_2 + (N-1)d$ with amplitude A_{2N+1}, and if there is no incident wave in region $y < y_1$, then $B_1 = 0$, B_{2N+1} represents the amplitude of the reflected wave and A_1 represents the amplitude of the transmitted wave, and, enforcing equation (3.33) with these values, the plane wave transmission coefficient results

$$t_N = \left.\frac{A_1}{A_{2N+1}}\right)_{B_1=0} = \frac{1}{T_{11}}. \tag{3.36}$$

If the medium of incidence in the half-space $y < y_1$ (medium I) is not of type A, as indicated in figure 3.7, but has constitutive parameters ε_I and μ_I, the reflection coefficient of the N equispaced layers is given by

$$r = \frac{r_{IA} + r_N}{1 + r_{IA} r_N}, \tag{3.37}$$

where r_N is given by equation (3.35) and r_{IA} is the amplitude reflection coefficient at the interface between medium I and A for waves impinging from the I side, given by equation (2.19). This expression can be derived easily from the Airy formula (2.30) by assuming the existence of an infinitesimally thin layer of medium A between the medium of incidence (in region $y < y_1 - \delta$) and the B layer and then taking the limit $\delta \to 0$.

3.4.3 1D photonic crystal

According to the Floquet–Bloch theorem [22], the fields for an ideal perfectly periodic lattice, with an infinite repetition along y of the basic inhomogeneity, are of the form

$$f(x, y) = f_{K_F}(x, y)e^{iK_F y}, \tag{3.38}$$

where the constant K_F is known as the Floquet–Bloch wavenumber and $f_{K_F}(x, y)$ is periodic along y (period d). For the AGBGA inhomogeneity, $d = d_A + d_B$, and in this case the phasors in equation (3.22) satisfy

$$f(x, y + d) = e^{ikx} f_{K_F}(y + d) e^{iK_F(y+d)} = e^{ikx} f_{K_F}(y) e^{iK_F y} e^{iK_F d} = f(x, y) e^{iK_F d} \quad (3.39)$$

Using this condition, the relation

$$A_3 e^{-i\beta_A y} e^{-i\beta_A d} + B_3 e^{i\beta_A y} e^{i\beta_A d} = A_1 e^{-i\beta_A y} e^{iK_F d} + B_1 e^{i\beta_A y} e^{iK_F d} \quad (3.40)$$

is obtained for amplitudes A_3 and B_3, corresponding to the fields in the region $y_2 < y < y_2 + d_A$, and amplitudes A_1 and B_1, corresponding to the fields in the region $y_1 - d_A < y < y_1$:

$$\begin{bmatrix} e^{-i\beta_A d} & 0 \\ 0 & e^{i\beta_A d} \end{bmatrix} \begin{bmatrix} A_3 \\ B_3 \end{bmatrix} = e^{iK_F d} \begin{bmatrix} A_1 \\ B_1 \end{bmatrix}. \quad (3.41)$$

Using equation (3.25), it follows that

$$\begin{bmatrix} e^{-i\beta_A d} & 0 \\ 0 & e^{i\beta_A d} \end{bmatrix} \mathbf{M} \begin{bmatrix} A_1 \\ B_1 \end{bmatrix} = e^{iK_F d} \begin{bmatrix} A_1 \\ B_1 \end{bmatrix}, \quad (3.42)$$

that is, the factor $e^{iK_F d}$ is an eigenvalue of the matrix $\mathbf{P}$

$$\mathbf{P} = \begin{bmatrix} e^{-i\beta_A d} & 0 \\ 0 & e^{i\beta_A d} \end{bmatrix} \mathbf{M} = \begin{bmatrix} P_{11} & P_{12} \\ P_{21} & P_{22} \end{bmatrix} \quad (3.43)$$

and the column vector with amplitudes A_1 and B_1 is the corresponding eigenvector. Note that the determinant of $\mathbf{P}$ and the determinant of $\mathbf{M}$ are both equal to one. Solving the characteristic equation of matrix $\mathbf{P}$, the eigenvalues $e^{iK_F d}$ are given by

$$e^{iK_F d} = \frac{(P_{11} + P_{22})}{2} \pm \sqrt{\left(\frac{P_{11} + P_{22}}{2}\right)^2 - 1}, \quad (3.44)$$

whereas the eigenvectors corresponding to these eigenvalues are

$$\begin{bmatrix} A_1 \\ B_1 \end{bmatrix} = \begin{bmatrix} P_{12} \\ e^{iK_F d} - P_{11} \end{bmatrix} = \begin{bmatrix} M_{12} e^{-i\beta_A d} \\ e^{iK_F d} - M_{11} e^{-i\beta_A d} \end{bmatrix}, \quad (3.45)$$

Taking into account the fact that the determinant of the similarity transformation of a matrix is equal to the determinant of the original matrix, we note that both solutions in equation (3.44) are the inverse of each other, that is, if K_F is one of the Floquet–Bloch wavenumbers, then $-K_F$ is the other. Adding both eigenvalues, we obtain the dispersion equation for the 1D photonic crystal

$$\cos K_F d = \frac{1}{2} (P_{11} + P_{22}) = \frac{1}{2} \text{tr}\{\mathbf{P}\}, \quad (3.46)$$

where tr denotes the trace of a matrix.

The semi-trace in equation (3.46) can be written as the sum of two terms, the usual term for a graphene-free photonic crystal (L_1) plus a term that becomes zero when $\sigma = 0$ (L_2). Using equations (3.26) and (3.43), and the expressions for matrices $\mathbf{M}_{A1}$, $\mathbf{M}_{A2}$, $\mathbf{M}_{B1}$ and $\mathbf{M}_{B2}$ for each polarization, it can be shown that the dispersion equation for the AGBGA unit cell photonic crystal takes the form [28]

$$\cos K_F d = L_1 + L_2, \tag{3.47}$$

The term corresponding to the binary system without graphene is

$$L_1 = \cos(\beta_B d_B) \cos(\beta_A d_A) - \frac{1}{2}\left(\frac{\eta_B \beta_A}{\eta_A \beta_B} + \frac{\eta_A \beta_B}{\eta_B \beta_A}\right)\sin(\beta_B d_B) \sin(\beta_A d_A), \tag{3.48}$$

where $\eta_j = \mu_j$ for s polarization or $\eta_j = \varepsilon_j$ for p polarization, and $j = A, B$. Using the dimensionless surface conductivity $\tilde{\sigma}$ defined in equation (1.41), the term that depends on σ can be written as

$$L_2 = -i4\pi\frac{\omega}{c}\alpha\tilde{\sigma}\left[\frac{\mu_A}{\beta_A}\sin(\beta_A d_A)\cos(\beta_B d_B) + \frac{\mu_B}{\beta_B}\sin(\beta_B d_B)\cos(\beta_A d_A)\right]$$
$$- 8\pi^2\alpha^2\frac{\omega^2}{c^2}\tilde{\sigma}^2\frac{\mu_A\mu_B}{\beta_A\beta_B}\sin(\beta_A d_A)\sin(\beta_B d_B) \tag{3.49}$$

for s polarization, and as

$$L_2 = -i4\pi\frac{c}{\omega}\alpha\tilde{\sigma}\left[\frac{\beta_A}{\varepsilon_A}\sin(\beta_A d_A)\cos(\beta_B d_B) + \frac{\beta_B}{\varepsilon_B}\sin(\beta_B d_B)\cos(\beta_A d_A)\right]$$
$$- 8\pi^2\alpha^2\frac{c^2}{\omega^2}\tilde{\sigma}^2\frac{\beta_A\beta_B}{\varepsilon_A\varepsilon_B}\sin(\beta_A d_A)\sin(\beta_B d_B) \tag{3.50}$$

for p polarization. When dissipation in the graphene sheets and in materials A and B in the unit cell is small enough to be ignored (ideal purely imaginary $\tilde{\sigma}$ and purely real η_j, $j = A, B$), the quantities L_1 and L_2 are both real-valued. Regimes where $|L_1 + L_2| < 1$ correspond to real values of K_F, that is, to propagating Floquet–Bloch waves (passbands or allowed bands), whereas regimes where $|L_1 + L_2| > 1$ correspond to values of K_F with a nonzero imaginary part, so the Floquet–Bloch waves are evanescent (stopbands or forbidden bands).

To illustrate how the inclusion of graphene sheets may change the transmission properties of 1D photonic crystals, we consider periodic AGBG multilayers with $d_A/d = 0.75$, $d_B/d = 0.25$, $d = 15$ μm, $\varepsilon_A = 2.76$, $\varepsilon_B = 7$ and $\mu_1 = \mu_2 = 1$, and graphene sheets with the Kubo parameters $T = 300$ K, $\mu_c = 0.25$ eV and $\hbar\gamma = 0.1$ meV. In figure 3.9 we compare the cases without (panel (a)) and with (panel (b)) graphene. In the vertical axis we show the frequency of the incident radiation in the spectral range between 0.25 and 16 THz, and in the horizontal axes we show the semi-trace $P/2$, equation (3.46), and the transmissivity at normal incidence for $N = 10$ unit cells. The graphene-free multilayer exhibits two regions where $|L_1| > 1$, centered around 5.24 and 10.48 THz. In these regions, corresponding to the usual Bragg gaps or forbidden bands of the perfectly periodic structure, the transmissivity of the structure with 10 unit cells is very low, indistinguishable from zero on the scales

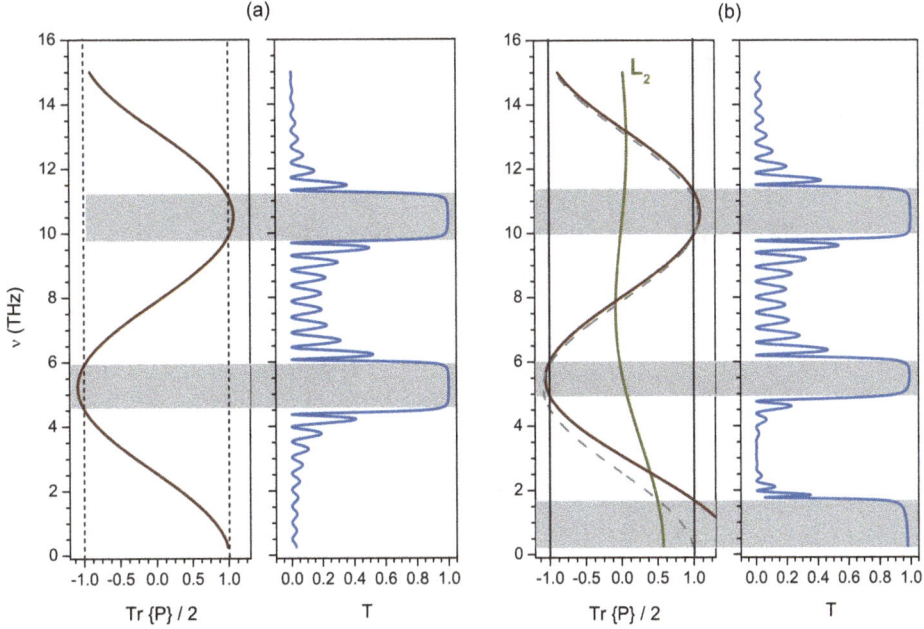

Figure 3.9. In the vertical axis the frequency of the incident radiation and in the horizontal axes the semi-trace of matrix $\mathbf{P}$ and the transmissivity at normal incidence for $N = 10$ unit cells for AGBG multilayers with $d_A/d = 0.75$, $d_B/d = 0.25$, $d = 15\,\mu m$, $\varepsilon_A = 2.76$, $\varepsilon_B = 7$, $\mu_1 = \mu_2 = 1$, and graphene layers with the Kubo parameters $T = 300$ K, $\mu_c = 0.25$ eV and $\hbar\gamma = 0.1$ meV. (a) Structure without graphene, $\mathrm{tr}\{\mathbf{P}\}/2 = L_1$. (b) Structure with graphene, $\mathrm{tr}\{\mathbf{P}\}/2 = L_1 + L_2$.

used. Due to the negligible values of $\tilde{\sigma}$ for frequencies near these gaps, $|L_2| \ll |L_1|$ (see curve for L_2 in figure 3.9(b)). Therefore, the position and widths of both gaps is only slightly altered by the presence of the graphene sheets, as can be observed in figure 3.9(b). However, due to the significant values of $\tilde{\sigma}$ in the low-frequency region (Im $\tilde{\sigma} > 10$ when $\nu < 1.92$ THz), a new gap, where $|L_1 + L_2| > 1$, appears for frequencies $\nu < 1.69$ THz. This new gap, completely originated by the presence of the graphene sheets, has been called *graphene-induced photonic band gap*. It exhibits photonic characteristics, such as omnidirectionality, polarization insensitivity and gate-voltage tunability, and is very different from conventional Bragg gaps [28]. In contrast to conventional graphene-free multilayers, the photonic band structure of periodic stacks of parallel graphene sheets does not disappear even when medium A is identical to medium B, that is, when the graphene sheets are immersed in a homogeneous medium. This can be appreciated in figure 3.10, which is completely analogous to figure 3.9, except that now $N = 20$ and medium A and medium B are both vacuum. In spite of the lack of optical contrast between media in the unit cell, we observe a graphene-induced photonic (non-Bragg) band gap for frequencies $\nu < 3.28$ THz, and even another gap centered around $\nu \approx 10.5$ THz. Oppositely to the graphene-induced gap, the occurrence of the gap around $\nu \approx 10.5$ THz is based on interference mechanisms between the waves scattered by the parallel graphene sheets, as evidenced by the fact that it corresponds to a wavelength very close to the

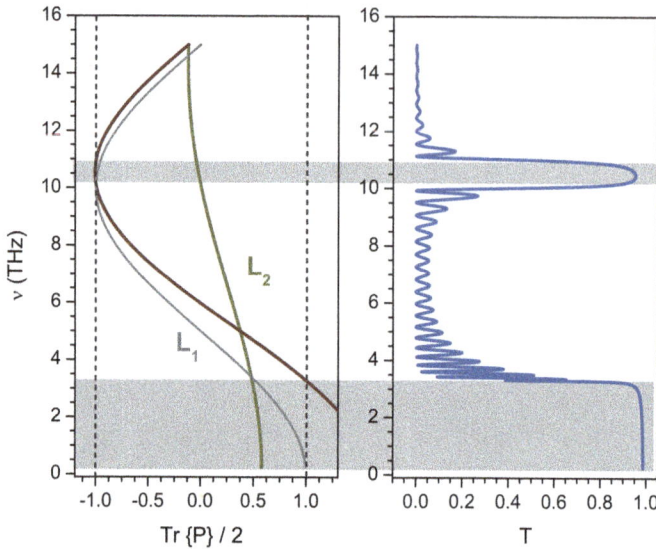

Figure 3.10. In the vertical axis the frequency of the incident radiation and in the horizontal axes the semi–trace $\text{tr}\{\mathbf{P}\}/2 = L_1 + L_2$ (left) and the transmissivity T (right) at normal incidence for the case $N = 20$, $\varepsilon_A = \varepsilon_B = \mu_1 = \mu_2 = 1$ (that is, parallel graphene sheets immersed in vacuum), $d_A/d = 0.75$, $d_B/d = 0.25$, $d = 15\,\mu\text{m}$, and Kubo parameters as in figure 3.9.

value $d = 30\,\mu\text{m}$ predicted by the Bragg formula for the first band of a unit cell with a unity average index of refraction. The Python script used to obtain the curves in figures 3.9 and 3.10 can be found in the appendix.

References

[1] Kretschmann E and Raether H 1968 Radiative decay of non-radiative surface plasmons excited by light *Z. Naturf.* A **23** 2135–6

[2] Otto A 1968 Excitation of nonradiative surface plasma waves in silver by the method of frustrated total reflection *Z. Physik.* **216** 398–410

[3] Barchiesi D and Otto A 2013 Excitations of surface plasmon polaritons by attenuated total reflection, revisited *Riv. Nuovo Cimento* **36** 173–209

[4] Bludov Y V, Vasilevskiy M and Peres N M R 2012 Tunable graphene-based polarizer *J. Appl. Phys.* **112** 084320

[5] de Oliveira R E P and de Matos C J S 2015 Graphene based waveguide polarizers: in-depth physical analysis and relevant parameters *Sci. Rep.* **5** 16949

[6] Sreekanth K V and Ting Y 2013 Long range surface plasmons in a symmetric graphene system with anisotropic dielectrics *J. Opt.* **15** 055002

[7] Bludov Y V, Vasilevskiy M I and Peres N M R 2010 Mechanism for graphene-based optoelectronic switches by tuning surface plasmon-polaritons in monolayer graphene *Eur. Phys. Lett.* **92** 68001

[8] Jiang-Tao L, Nian-Hua L L W, Xin-Hua D and Fu-Hai S 2013 Gate-tunable nearly total absorption in graphene with resonant metal back reflector *Eur. Phys. Lett.* **104** 57002

[9] Fang-Fang Y, Ying-Long H, Wen-Bo X, Jiang-Tao L and Nian-Hua L 2015 Control of absorption of monolayer MoS_2 thin-film transistor in one-dimensional defective photonic crystal *Eur. Phys. Lett.* **112** 37008

[10] Liu F, Chong Y D, Adam S and Polini M 2014 Gate-tunable coherent perfect absorption of terahertz radiation in graphene *2D Materials* **1** 031001

[11] Born M and Wolf E 1986 *Principles of Optics* 6th edn (Oxford: Pergamon) section 7.6.1

[12] Mason D R, Menabde S G and Park N 2014 Unusual otto excitation dynamics and enhanced coupling of light to TE plasmons in graphene *Opt. Express* **22** 847–58

[13] Hass J, Varchon F, Millán-Otoya J E, Sprinkle M, Sharma N, de Heer W A, Berger C, First P N, Magaud L and Conrad E H 2008 Why multilayer graphene on 4H-SiC(0001) behaves like a single sheet of graphene *Phys. Rev. Lett.* **100** 125504

[14] Cuevas M and Depine R A 2008 Excitation of surface plasmon polaritons along the sinusoidal boundary of a metamaterial *Phys. Rev.* B **78** 125412

[15] Foley J J, Harutyunyan H, Rosenmann D, Divan R, Wiederrecht G P and Gray S K 2015 When are surface plasmon polaritons excited in the kretschmann-raether configuration? *Sci. Rep.* **5** 9929

[16] Zeller M, Cuevas M and Depine R A 2011 Surface plasmon polaritons in attenuated total reflection systems with metamaterials: homogeneous problem *J. Opt. Soc. Am.* B **28** 2042–7

[17] Zeller M, Cuevas M and Depine R A 2012 Phase and reflectivity behavior near the excitation of surface plasmon polaritons in Kretschmann-ATR systems with metamaterials *Eur. Phys. J.* D **66** 1–7

[18] Yablonovitch E 1987 Inhibited spontaneous emission in solid-state physics and electronics 1987 *Phys. Rev. Lett.* **58** 2059–62

[19] John S 1987 Strong localization of photons in certain disordered dielectric superlattices 1987 *Phys. Rev. Lett.* **58** 2486–9

[20] Felbacq D and Zolla F 2003 *Introduction to complex mediums for optics and electromagnetics* ed W S Weiglhofer and A Lakhtakia (Bellingham, WA: SPIE) pp 365–93

[21] Haus J W 2004 *Nanometer Structures: Theory, Modeling and Simulation* ed A Lakhtakia (Bellingham, WA: SPIE) pp 45–106 ch 3

[22] Yeh P 1988 *Optical Waves in Layered Media* (New York: Wiley)

[23] Wu L, He S and Shen L 2003 Band structure for a one-dimensional photonic crystal containing left-handed materials *Phys. Rev.* B **67** 235103

[24] Li J, Zhou L, Chan C T and Sheng P 2003 Photonic band gap from a stack of positive and negative index materials *Phys. Rev. Lett.* **90** 083901

[25] Depine R A, Martínez-Ricci M L, Monsoriu J A, Silvestre E and Andrés P 2007 Zero permeability and zero permittivity band gaps in 1D metamaterial photonic crystals *Phys. Lett.* A **364** 352–5

[26] Monsoriu J A, Depine R A, Martínez-Ricci M L and Silvestre E 2006 Interaction between non-Bragg band gaps in 1D metamaterial photonic crystals *Opt. Express* A **14** 12958–67

[27] Gómez A, Martínez-Ricci M L, Depine R A and Lakhtakia A 2009 Photonic band gap materials comprising positive-phase-velocity and negative-phase-velocity layers in waveguides *J. Modern Opt.* **56** 1688–97

[28] Madani A and Entezar S R 2013 Optical properties of one-dimensional photonic crystals containing graphene sheets *Phys. B Condens. Matter* **431** 1–5

[29] Bludov Y V, Peres N M R and Vasilevskiy M I 2013 Unusual reflection of electromagnetic radiation from a stack of graphene layers at oblique incidence *J. Opt.* **15** 114004

[30] Zhan T, Shi X, Dai Y, Liu X and Zi J 2013 Transfer matrix method for optics in graphene layers *J. Phys. Condens. Matter* **25** 215301

[31] Zhang Y, Wu Z, Cao Y and Zhang H 2015 Optical properties of one-dimensional fibonacci quasi-periodic graphene photonic crystal *Opt. Commun.* **338** 168–73

[32] Yan H, Li X, Chandra B, Tulevski G, Wu Y, Freitag M, Zhu W, Avouris P and Xia F 2012 Tunable infrared plasmonic devices using graphene/insulator stacks *Nat. Nanotechnol.* **7** 330–4

[33] Baek I H, Ahn K J, Kang B J, Bae S, Hong B H, Yeom D, Lee K, Jeong Y U and Rotermund F 2013 Terahertz transmission and sheet conductivity of randomly stacked multi-layer graphene *Appl. Phys. Lett.* **102** 191109

[34] Othman M A K, Guclu C and Capolino F 2013 Graphene-dielectric composite metamaterials: evolution from elliptic to hyperbolic wavevector dispersion and the transverse epsilon-near-zero condition *J. Nanophotonics* **7** 073089

[35] Sernelius B E 2014 Electromagnetic normal modes and casimir effects in layered structures *Phys. Rev. B* **90** 155457

[36] Griffiths D J and Steinkea C A 2001 Waves in locally periodic media *Am. J. Phys.* **69** 137–54

Chapter 4

Graphene gratings

In this chapter we shall relax the assumption of perfectly flat graphene layers characterized by a surface conductivity that does not vary along the layer and consider the case of geometric and constitutive *periodic variations*, that is, graphene gratings. These variations offer more degrees of freedom in the manipulation of infrared and THz electromagnetic waves. There are many potential applications, including antennas and absorbers, fixed and reconfigurable frequency selective surfaces, periodic plasmonic waveguides, tunable switches, and infrared and THz sensors [1–10]. Constitutive modulations can be formed without patterning either the graphene sheet or the dielectric substrate, and they can be induced by a periodic strain field using absorbed molecules with a modulated adsorption profile, or by modulating the electron concentration by inducing a periodic potential [11–13].

Having described the basics of phase matching-devices that use ATR to excite propagating graphene plasmons in chapter 3, we continue the discussion by providing a number of examples of *grating coupling*, another phase-matching mechanism which allows the excitation of graphene plasmons via direct illumination, and is based on the use of some kind of diffraction grating that changes the parallel wave vector component of incident radiation by the addition or subtraction of an integer number of the grating wavevector $2\pi/d$, d the period. While large wavenumber mismatches between graphene surface plasmons and free-space radiation can be advantageous for many applications because they provide high electromagnetic confinement, they also make graphene surface plasmons difficult to excite using ATR couplers like those discussed in section 3.3, because a large wavenumber mismatch could only be compensated with exotic or unavailable materials with a high index of refraction.

The excitation of SP modes in periodic ultrathin metallic structures has been experimentally and theoretically studied for both one-dimensional (1D) and two-dimensional (2D) arrays [14]. Similar studies for the 2D limit, i.e. for a graphene monoatomic layer, are therefore a natural continuation of this research. The

doi:10.1088/978-1-6817-4309-7ch4

analytical solution to the problem of scattering of electromagnetic radiation by a square-wave grating with a flat graphene sheet on top was given in [15]. In particular, plasmon excitations in patterned graphene structures have been explored to increase optical absorption in graphene due to the optical-field enhancement occurring at these resonances. The excitation of graphene plasmons on 1D micro-ribbon gratings was demonstrated experimentally in [16], and systematic theoretical studies for these structures over a wide range of parameters were given in [17]. Light–plasmon coupling in graphene disks arranged in a triangular lattice was demonstrated experimentally in [18], and the possibility of having total light absorption at a periodic array of graphene nanodisks was studied theoretically in [19]. In addition to periodically patterned graphene structures, optical absorption enhancement without the need for engineering graphene was demonstrated theoretically [20] using a graphene sheet placed on top of dielectric gratings consisting of either 1D dielectric strips or a 2D lattice of dielectric cylinders in a dielectric background.

In this chapter we give a basic introduction to the electromagnetic theory of graphene gratings. We begin by formulating the boundary value problem for gratings that combine both conductivity and interface variations. After introducing the Rayleigh expansions in terms of Floquet–Bloch space harmonics, we show how to calculate optical quantities of interest, such as the diffraction efficiency or graphene absorption. Then, the mode-matching technique is applied to two kinds of flat gratings. First, it is applied to a periodic array of parallel graphene strips lying on a flat substrate, a structure that allows us to introduce a geometrical model for the excitation of localized graphene surface plasmons. Second, it is applied to a sinusoidally modulated conductivity grating supporting propagating surface plasmons, which are closely related to the propagating surface plasmons described in the previous chapters.

4.1 Boundary value problem

In figure 4.1 we sketch a 1D graphene grating that combines both conductivity and interface variations. Due to the presence of graphene, the boundary

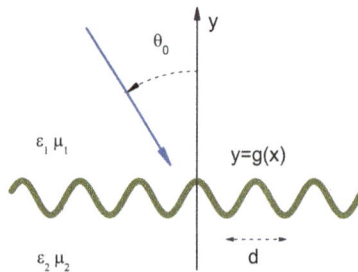

Figure 4.1. Schematic of the boundary value problem for a graphene grating combining both conductivity and interface variations. A plane wave is incident at an angle θ_0 with respect to the y axis on periodically corrugated boundary $y = g(x)$ (period d), which is tightly coated with a graphene mantle characterized by a spatially varying surface conductivity. The corrugated boundary separates two homogeneous, isotropic and linear dielectric–magnetic materials.

$y = g(x) = g(x + d)$ exhibits a surface conductivity $\sigma u(x)$, with $u(x)$ being a periodic function that modulates the conductivity. For a homogeneously doped graphene sheet that is tightly adhered to the corrugation, $u(x) = 1$ and the conductivity does not depend on x, whereas for a periodic array of parallel graphene strips lying on a flat substrate, $g(x) = 0$ and $u(x) = 1$ on the strips and 0 in the space between strips. The grooves of the grating are parallel to the z axis and the boundary separates two homogeneous, isotropic, linear and passive dielectric–magnetic materials characterized by the constitutive parameters ε_1 and μ_1 ($y > 0$, medium of incidence) and ε_2 and μ_2 ($y < 0$, medium of transmission).

4.1.1 Field representation

A linearly polarized electromagnetic plane wave is incident on the graphene grating at an angle θ_0 with respect to the y axis. This plane wave can be either s-polarized or p-polarized and, given that its propagation direction is contained in the xy plane, all reflected and transmitted fields must be linearly polarized in the same way as the incident plane wave [21], while their propagation directions are also contained in the xy plane.

The function $f(x, y)$ representing the z-directed component of the total electric field for the s polarization case and the z-directed component of the total magnetic field for the p polarization case, satisfies Helmholtz equations of the type (1.45) in the media of incidence and transmission. In the medium of incidence, $f(x, y)$ can be represented as an incident field plus a superposition of diffracted–reflected plane waves in the form of space harmonics [22–25]:

$$f(x, y) = \exp\left[i\left(\alpha_0 x - \beta_0^{(1)} y\right)\right] + \sum_{n=-\infty}^{\infty} c_n^{(1)} \exp\left[i\left(\alpha_n x + \beta_n^{(1)} y\right)\right]. \tag{4.1}$$

A similar superposition of diffracted–transmitted space harmonics can be used to represent the total field in the medium of transmission [22–25]:

$$f(x, y) = \sum_{n=-\infty}^{\infty} c_n^{(2)} \exp\left[i\left(\alpha_n x - \beta_n^{(2)} y\right)\right]. \tag{4.2}$$

The complex amplitudes $c_n^{(p)}$, ($-\infty < n < \infty$; $p = 1, 2$), are scalar coefficients that are determined by the solution of a boundary value problem, while

$$\alpha_n = \frac{\omega}{c} \sin \theta_0 + 2n\pi/d$$

$$\beta_n^{(p)} = \sqrt{\frac{\omega^2}{c^2} \varepsilon_p \mu_p - \alpha_n^2} \quad p = 1, 2. \tag{4.3}$$

Due to the presence of the square roots, both $\beta_n^{(1)}$ and $\beta_n^{(2)}$ are double-valued functions. If $\beta_n^{(p)}$ represents an *up-going* plane wave, then $-\beta_n^{(p)}$ represents a *down-going* wave, and vice versa. We must choose the correct signs for all $\beta_n^{(1)}$ (up-going plane waves), as well as for all $\beta_n^{(2)}$ (down-going plane waves). If $\beta_n^{(p)}$ is real-valued, it corresponds to a propagating plane wave, otherwise it corresponds to an evanescent wave. On the assumption that both the medium of incidence and the medium of transmission are lossless media, $\beta_n^{(p)}$, $p = 1, 2$, they are either purely real or purely imaginary. Therefore, for positive-index media, the condition

$$\left.\begin{array}{l} \mathrm{Re}\,\beta_n^{(p)} \geqslant 0 \\ \mathrm{Im}\,\beta_n^{(p)} \geqslant 0 \end{array}\right\} \forall n \qquad (4.4)$$

must hold to obtain up-going plane waves in the upper half-space ($y > 0, p = 1$) and down-going plane waves in the lower half-space ($y < 0, p = 2$).

When the angle of incidence or the angular frequency of the incident plane wave changes, $\beta_n^{(p)}$ may change from purely real to purely imaginary, or vice versa. Such alterations, corresponding to a diffracted–reflected or a diffracted–transmitted space harmonic just grazing the surface of the grating, are usually marked by noticeable discontinuities in the diffraction spectra and hence they are called *Rayleigh–Wood anomalies* [22–25].

If the graphene strip grating is deposited on the surface of a passive lossy medium, Im $\varepsilon_2 > 0$ and Im $\mu_2 > 0$, the diffracted–transmitted space harmonics must attenuate as $y \to -\infty$. Therefore,

$$\mathrm{Im}\,\beta_n^{(2)} > 0 \quad \forall n, \qquad (4.5)$$

and this condition on Im $\beta_n^{(2)}$ automatically fixes the sign of Re $\beta_n^{(2)}$, regardless of whether the lossy material has a positive or a negative refractive index [26–28].

4.1.2 Boundary conditions

Analogously to the reflection and transmission problems considered previously, boundary conditions (1.50) and (1.51) must be invoked to find the amplitudes of the diffracted–reflected and diffracted–transmitted space harmonics ($c_n^{(1)}$ and $c_n^{(2)}$). However, the problem for the graphene gratings considered here exhibits two main differences with respect to the problems considered in chapters 2 and 3: (i) the boundary is no longer perfectly flat, or (ii) the graphene layer is no longer characterized by a surface conductivity that is constant along the layer. The separate or simultaneous appearance of these changes makes the mathematical description and numerical computation of the diffracted reflected and transmitted fields more complicated than in cases involving flat, homogeneous interfaces, as will be seen in the examples considered in the following sections.

In terms of $f(x, y)$, the boundary conditions on each period of the graphene grating can be written as

$$\left.\begin{array}{l} f_1(x, g(x)) = f_2(x, g(x)) \\ \dfrac{1}{\mu_2}\dfrac{\partial f_2}{\partial \mathbf{n}}(x, g(x)) - \dfrac{1}{\mu_1}\dfrac{\partial f_1}{\partial \mathbf{n}}(x, g(x)) = i\omega\mu_0\sigma u(x)f_2(x, g(x)) \end{array}\right\} \ s \text{ polarization} \qquad (4.6)$$

and

$$\left.\begin{array}{l} \dfrac{1}{\varepsilon_1}\dfrac{\partial f_1}{\partial n}(x, g(x)) = \dfrac{1}{\varepsilon_2}\dfrac{\partial f_2}{\partial n}(x, g(x)) \\[2em] f_2(x, g(x)) - f_1(x, g(x)) = -i\dfrac{\sigma\; u(x)}{\omega\varepsilon_0}\dfrac{1}{\varepsilon_2}\dfrac{\partial f_2}{\partial n}(x, g(x)) \end{array}\right\} \quad p \text{ polarization,} \qquad (4.7)$$

where n is a unit vector pointing in the direction of the normal to the boundary $y = g(x)$ and oriented from medium 2 to medium 1. These boundary conditions have almost the same form as equations (2.6) and (2.11), except that (i) the derivative with respect to y has been replaced by a normal derivative, (ii) the fields are evaluated on $y = g(x)$ instead of $y = 0$ and (iii) the constant σ has been replaced by the value $\sigma\; u(x)$, which can vary along a grating period.

Only outgoing waves are included in expansions (4.1) and (4.2) for the diffracted field above and below the grating interface. These expansions have the same form as those used by Rayleigh [29] when considering the scattering from a sinusoidally undulated interface and are known as *Rayleigh expansions* [30]. Whereas the validity of Rayleigh expansions is unquestionable outside the corrugation region $\max g(x) > y > \min g(x)$, their validity inside the corrugation region has been questioned [23] on the grounds that the diffracted field in this region should actually consist of both outgoing and incoming waves. Therefore, the use of expansions (4.1) and (4.2) in the boundary conditions, the procedure originally used by Rayleigh and known as the *Rayleigh hypothesis*, has long been considered by the diffraction community to be a valid approximation when corrugations are not deep, although unexpected agreement between the results given by the Rayleigh method and those obtained using methods considered to be exact has been mentioned in the literature [31, 32] a few times. Recent numerical experiments seems to indicate that the convergence problems in applying the Rayleigh method are only caused by the limited precision of the computer arithmetic being used and that increasing processor precision improves the convergence and finally leads to exact results, whatever the groove depth and prescribed accuracy [30, 31]. However, when arbitrary-precision arithmetic is not an option, the Rayleigh hypothesis is avoided or is only used for very shallow corrugations, as is the case for the relief graphene gratings considered in [1, 12].

4.1.3 Diffraction efficiencies and graphene absorption

Under the assumption of lossless media, diffraction efficiencies are defined for the reflected and transmitted diffracted orders as the ratio between the power carried to infinity by a diffracted order and the power incident onto a period of the grating. Using the complex Poynting vector to calculate time-averaged power flows, the efficiencies are given by

$$e_n^r = \operatorname{Re}\frac{\beta_n^{(1)}}{\beta_0^{(1)}}\left|c_n^{(1)}\right|^2 \qquad (4.8)$$

for the reflected orders, and by

$$e_n^t = \frac{\eta_1}{\eta_2} \, \text{Re} \, \frac{\beta_n^{(2)}}{\beta_0^{(1)}} \left| c_n^{(2)} \right|^2 \tag{4.9}$$

for the transmitted orders, where $\eta = \mu$ for s polarization and $\eta = \varepsilon$ for p polarization.

The total power diffracted in the far field normalized to the incident power results in

$$P_d = P_{dr} + P_{dt}, \tag{4.10}$$

where

$$P_{dr} = \sum_{n \in U^{(1)}} e_n^r \tag{4.11}$$

represents the fraction of the incident power that is reflected back into the medium of incidence, whereas

$$P_{dt} = \sum_{n \in U^{(2)}} \text{Re} \, e_n^t \tag{4.12}$$

represents the fraction of the incident power that is transmitted into the medium of transmission. The sets $U^{(p)}$, $p = 1, 2$, in equations (4.11) and (4.12) contain the indexes corresponding to propagating plane waves in medium p, that is, to space harmonics with $\text{Im} \beta_n^{(p)} = 0$. Using the conservation of power principle, the normalized power absorbed per period at the graphene mantle located at $y = g(x)$ is given by

$$P_a = 1 - P_d = 1 - \sum_{n \in U^{(1)}} \text{Re} \, \frac{\beta_n^{(1)}}{\beta_0^{(1)}} \left| c_n^{(1)} \right|^2 - \sum_{n \in U^{(2)}} \text{Re} \, \frac{\eta_1 \beta_n^{(2)}}{\eta_2 \beta_0^{(1)}} \left| c_n^{(2)} \right|^2, \tag{4.13}$$

that is, when the spaces above and below the grating boundary are lossless media, the part of the incident radiation that is neither reflected nor transmitted is absorbed by graphene. If, instead of a lossless transmission media, the region $y < g(x)$ is occupied by a lossy dielectric or metal, the power carried to infinity by any diffracted–transmitted order is zero, and then P_{dt} in equation (4.12) equals zero and the quantity P_a now has two contributions

$$P_a = P_{ag} + P_{at}. \tag{4.14}$$

The first term is a surface contribution that comes from the dissipation of electric currents in the graphene mantle, whereas the second is a volume contribution from the loss in the medium of transmission. P_{ag} is calculated using the time-averaged power dP_{ag}/da transferred per unit surface from the electromagnetic fields to the graphene charge carriers. This quantity, given by equation (1.33), involves the tangential component of the electric field, which is continuous at $y = g(x)$ (see equations (4.6) and (4.7)). To obtain P_{ag} we integrate dP_{ag}/da over a period and divide the result by the power incident per period. Thus, the second term in

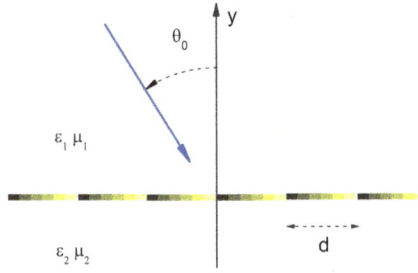

Figure 4.2. Conductivity grating. A flat graphene sheet, characterized by a spatially varying surface conductivity, is lying on the flat interface between two homogeneous, isotropic and linear dielectric–magnetic materials.

equation (4.14), which gives the power dissipated in the medium of transmission, can be obtained as

$$P_{at} = 1 - P_d - P_{ag}, \tag{4.15}$$

that is, the part of the incident radiation that is neither diffracted towards $y \to \infty$ nor absorbed by the graphene sheet is dissipated in the transmission medium. Alternatively, the normalized power dissipated in the transmission medium can be obtained by calculating the power flux through a surface just below the graphene mantle.

4.2 Flat graphene gratings

Very simple flat graphene gratings can be obtained by depositing geometrically patterned graphene structures or a single periodically doped graphene sheet (see figure 4.2) on the non-corrugated boundary between two media. From a theoretical point of view, this case has the advantage that the Rayleigh hypothesis is unquestionable and the Rayleigh expansions (4.1) and (4.2) can be safely used in the boundary conditions. In this case, $g(x) = 0$, and equations (4.6) and (4.7) adopt the simpler form

$$\left. \begin{aligned} f_1(x, 0) &= f_2(x, 0) \\ \frac{1}{\mu_2} \frac{\partial f_2}{\partial y}(x, 0) - \frac{1}{\mu_1} \frac{\partial f_1}{\partial y}(x, 0) &= i\omega\mu_0 \sigma u(x) f_2(x, 0) \end{aligned} \right\} \quad s \text{ polarization} \tag{4.16}$$

and

$$\left. \begin{aligned} \frac{1}{\varepsilon_1} \frac{\partial f_1}{\partial y}(x, 0) &= \frac{1}{\varepsilon_2} \frac{\partial f_2}{\partial y}(x, 0) \\ f_2(x, 0) - f_1(x, 0) &= -i\frac{\sigma u(x)}{\omega\varepsilon_0} \frac{1}{\varepsilon_2} \frac{\partial f_2}{\partial y}(x, 0) \end{aligned} \right\} \quad p \text{ polarization.} \tag{4.17}$$

Except for the additional presence of the modulation function $u(x)$ in the equations containing σ, the boundary conditions (4.16) and (4.17) are almost identical to those obtained in section 2.1 for the problem of the reflection and transmission of an electromagnetic wave at a planar graphene sheet. In fact, when the modulation disappears, that is, when $u(x) = 1$, the diffraction problem considered here becomes identical to the problem considered in section 2.1. In that case, only the space harmonics with $n = 0$ in the expansions of the diffracted–reflected and diffracted–transmitted fields in equations (4.1) and (4.2) are necessary to satisfy the boundary conditions at $y = 0$, and therefore only the amplitudes $c_0^{(1)}$ and $c_0^{(2)}$ are different from zero when the function $u(x)$ in equations (4.16) and (4.17) is a constant. On the other hand, for a true conductivity grating, $u(x)$ is not a constant and in principle all the space harmonics in expansions (4.1) and (4.2) are necessary to satisfy the boundary conditions at $y = 0$.

A variety of numerical techniques can be used to find the scalar coefficients $c_n^{(p)}$ for flat graphene gratings [1, 6, 7, 10, 17]. Here we give a detailed presentation of the mode-matching method that uses Fourier expansion. Although in some cases it may exhibit slow convergence [1, 7, 10, 33]—particularly for some rather 'pathological' modulation functions $u(x)$, such as the rectangular function corresponding to strip gratings—the method provides a physically based mathematical formulation and facilitates understanding of the wave processes involved in graphene gratings.

The first step in every grating mode-matching method is to obtain the modes, that is, the set of well-defined solutions to the Maxwell equations that satisfies all the physical requirements except the boundary conditions at the surface of the grating. For flat gratings such as those considered here, the modal solutions are just the set of space harmonics appearing in equations (4.1) and (4.2). For metallic curved strips, different modal solutions—more efficient in fitting the geometry of the problem—have been obtained [25]. The second step in the mode-matching method is to determine the modal amplitudes ($c_n^{(p)}$) from the boundary conditions. These modal amplitudes are all that is needed to calculate the electromagnetic quantities of interest, such as diffraction efficiencies, surface currents or graphene absorption. To match the modes at the surface of the grating, it is convenient to separate the analysis for each polarization case.

4.2.1 s polarization

After substitution of equations (4.1) and (4.2) into equation (4.16), we obtain the two equations below

$$\sum_{n=-\infty}^{\infty} \left[-c_n^{(1)} + c_n^{(2)} \right] \exp(i\alpha_n x) = \exp(i\alpha_0 x) \tag{4.18}$$

$$\sum_{n=-\infty}^{\infty} \left[\frac{\beta_n^{(1)}}{\mu_1} c_n^{(1)} + \left(\frac{\beta_n^{(2)}}{\mu_2} + \omega\mu_0\sigma u(x) \right) c_n^{(2)} \right] \exp(i\alpha_n x) = \frac{\beta_0^{(1)}}{\mu_1} \exp(i\alpha_0 x). \tag{4.19}$$

Multiplying both sides of equation (4.18) by $\exp(i\alpha_m x)$, $-\infty < m < \infty$, and taking the integral over the interval $0 < x < d$ and using the orthogonality of the space harmonics, we obtain

$$-c_m^{(1)} + c_m^{(2)} = \delta_{m0}, \tag{4.20}$$

where δ_{m0} represents the Kronecker delta. Repeating the same procedure for equation (4.19) and using the definition (1.41) for the dimensionless surface conductivity $\tilde{\sigma}$, we find

$$\frac{\beta_m^{(1)}}{\mu_1}c_m^{(1)} + \sum_{n=-\infty}^{\infty}\left(\frac{\beta_m^{(2)}}{\mu_2}\delta_{mn} + 4\pi\alpha\tilde{\sigma}\frac{\omega}{c}L(m-n)\right)c_n^{(2)} = \frac{\beta_0^{(1)}}{\mu_1}\delta_{m0}, \qquad (4.21)$$

where

$$L(p) = \frac{1}{d}\int_0^d u(x)\exp(-i2\pi px/d)\mathrm{d}x. \qquad (4.22)$$

Multiplying equation (4.20) by $\beta_m^{(1)}/\mu_1$ and adding the result to equation (4.21), we finally obtain an infinite system of linear equations for the amplitudes of the diffracted–transmitted space harmonics:

$$\sum_{n=-\infty}^{\infty}\left[\left(\frac{\beta_m^{(1)}}{\mu_1} + \frac{\beta_m^{(2)}}{\mu_2}\right)\delta_{mn} + 4\pi\alpha\tilde{\sigma}\frac{\omega}{c}L(m-n)\right]c_n^{(2)} = 2\frac{\beta_0^{(1)}}{\mu_1}\delta_{m0}. \qquad (4.23)$$

4.2.2 p polarization

In the case of p polarization we place the field expansions given by equations (4.1) and (4.2) into the boundary conditions given by equations (4.17). Doing this, we now obtain

$$\sum_{n=-\infty}^{\infty}\left[\frac{\beta_n^{(1)}}{\varepsilon_1}c_n^{(1)} + \frac{\beta_n^{(2)}}{\varepsilon_2}c_n^{(2)}\right]\exp(i\alpha_n x) = \frac{\beta_0^{(1)}}{\varepsilon_1}\exp(i\alpha_0 x) \qquad (4.24)$$

$$\sum_{n=-\infty}^{\infty}\left[-c_n^{(1)} + \left(1 + \frac{\sigma}{\omega\varepsilon_0}\frac{\beta_n^{(2)}}{\varepsilon_2}u(x)\right)c_n^{(2)}\right]\exp(i\alpha_n x) = \exp(i\alpha_0 x). \qquad (4.25)$$

Multiplying both sides of these equations by $\exp(i\alpha_m x)$, $-\infty < m < \infty$, and taking the integral over the interval $0 < x < d$ and using the orthogonality of the space harmonics, we obtain an infinite set of coupled equations for the amplitudes of the reflected and transmitted space harmonics:

$$\frac{\beta_m^{(1)}}{\varepsilon_1}c_m^{(1)} + \frac{\beta_m^{(2)}}{\varepsilon_2}c_m^{(2)} = \frac{\beta_0^{(1)}}{\varepsilon_1}\delta_{m0} \qquad (4.26)$$

$$-c_m^{(1)} + \sum_{n=-\infty}^{\infty}\left(\delta_{mn} + \frac{\beta_m^{(2)}}{\varepsilon_2}4\pi\alpha\tilde{\sigma}\frac{c}{\omega}L(m-n)\right)c_n^{(2)} = \delta_{m0}. \qquad (4.27)$$

Multiplying equation (4.27) by $\beta_m^{(1)}/\varepsilon_1$ and adding the result to equation (4.26), we arrive at an infinite system of linear equations for amplitudes $c_n^{(2)}$ alone:

$$\sum_{n=-\infty}^{\infty} \left[\left(\frac{\beta_m^{(1)}}{\varepsilon_1} + \frac{\beta_m^{(2)}}{\varepsilon_2} \right) \delta_{mn} + \frac{\beta_m^{(1)}}{\varepsilon_1} \frac{\beta_m^{(2)}}{\varepsilon_2} 4\pi\alpha\sigma \frac{\omega}{c} L(m-n) \right] c_n^{(2)} = 2\frac{\beta_0^{(1)}}{\varepsilon_1} \delta_{m0}. \qquad (4.28)$$

4.3 Strip gratings

Strip gratings (see figure 4.3) are easy to manufacture and simple to describe theoretically. They consist of periodical arrays of parallel identical coplanar graphene strips of width $w < d$ deposited on the interface between two homogeneous, isotropic and linear dielectric–magnetic materials. The electromagnetic response of these structures has been studied using the Fourier expansion method [7, 17] with a singular integral equations method [6] and the analytical regularization technique [10]. To use the mode-matching method for flat conductivity gratings described in the previous section, we set $u(x) = 1$ for $0 < x < d$ and $u(x) = 0$ for $w < x < d$. In this case, the coefficients $L(p)$ in equation (4.22) are given analytically as

$$L(p) = \begin{cases} [\exp(iq2\pi w/d) - 1]/(iq2\pi) & q \neq 0 \\ w/d & q = 0. \end{cases} \qquad (4.29)$$

4.3.1 Removing strips from a graphene sheet

We already know (see equation (2.29)) that a normally illuminated free-standing, unpatterned flat graphene sheet in vacuum has a polarization-independent, near-unity transmittance T and an absorptance that is given approximately by $A \approx (1 - T)$. Considering that such a flat sheet can be thought of as a graphene-strip grating with $w = d$, it could be argued that true graphene-strip gratings with

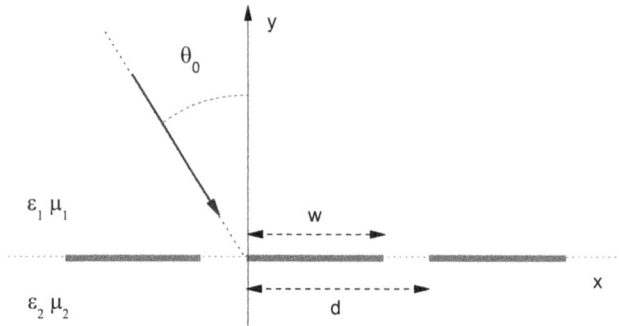

Figure 4.3. Strip grating. A plane wave illuminates a periodical array (period d) of parallel coplanar graphene strips (width w) lying on the interface between two homogeneous, isotropic and linear dielectric–magnetic materials.

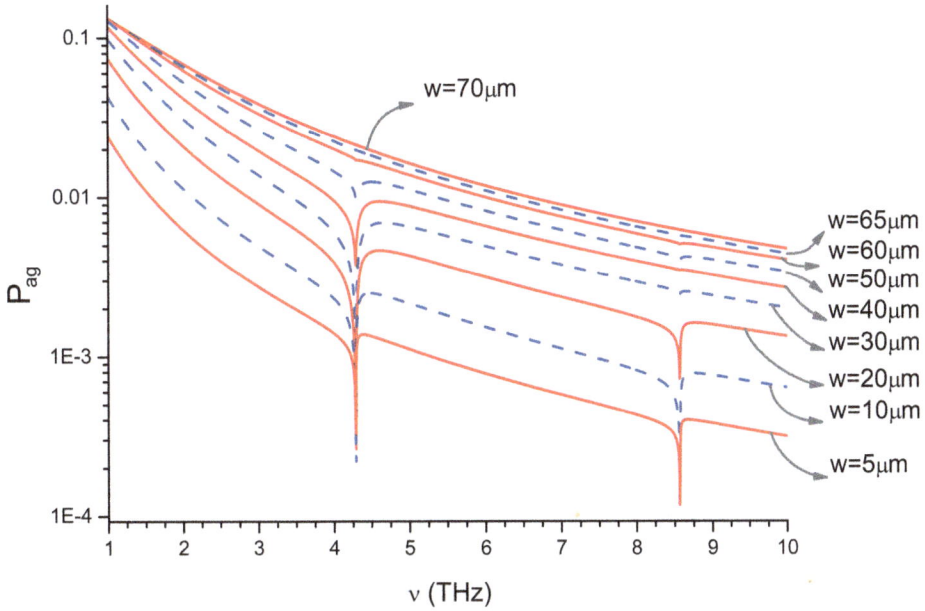

Figure 4.4. P_a (equation (4.13)) as a function of the frequency for free-standing graphene-strip gratings with $d = 70$ μm and different strip widths of $w = 5, 10, 20, 30, 40, 50, 60$ and 65 μm. The grating is normally illuminated by an s-polarized incident plane wave, $\varepsilon_1 = \varepsilon_2 = 1$, $\mu_1 = \mu_2 = 1$, $\mu_c = 0.39$ eV, $T = 300$ K and $\hbar\gamma = 0.658$ meV. The absorption of an unpatterned graphene sheet ($w = d = 70$ μm) is given as a reference. The Python script used to obtain these curves is given in the appendix.

$w < d$, with less graphene per period, should have lower absorption and higher transmittance than the unpatterned flat graphene sheet. For s polarization, the curves of P_a (equation (4.13)) shown in figure 4.4 confirm this expectation. These curves correspond to normally illuminated free-standing graphene-strip gratings with period $d = 70$ μm and different values for the strip width w, ranging from $w = 5$ μm, that is, a grating with very narrow graphene strips, to $w = 65$ μm, that is, a graphene sheet with very narrow holes. When $w/d \to 1$, the curves of P_a as a function of frequency tend to the curve corresponding to the fraction of incident power that is absorbed by the unpatterned graphene sheet, whereas lower absorption (and higher transmittance, see figure 4.5) is obtained when $w/d \to 0$. In normal incidence, this grating exhibits Rayleigh–Wood anomalies near $\nu \approx 4.28$ THz (where $\lambda = d$) and $\nu \approx 8.56$ THz (where $\lambda = d/2$). For this polarization, the anomalies are accompanied by decreases in P_a (figure 4.4) and increases in the total transmissivity ($\approx 1 - P_a$). When $\nu < 4.28$ THz, there is only one propagating wave (the space harmonic with $n = 0$) in both the reflected–diffracted and the transmitted–diffracted fields, and therefore the total transmissivity in this region is equal to the efficiency e_0^t of the 0th transmitted–diffracted order shown in figure 4.5. However, in the range 4.28 THz $< \nu < 8.56$ THz, three space harmonics (those with $-1 \leqslant n \leqslant 1$) contribute to the far field in the half-spaces above and below the graphene strips and, therefore, due to the normal incidence symmetry condition, the total transmissivity differs from e_0^t by twice the value of the efficiency of the +1th transmitted–diffracted order.

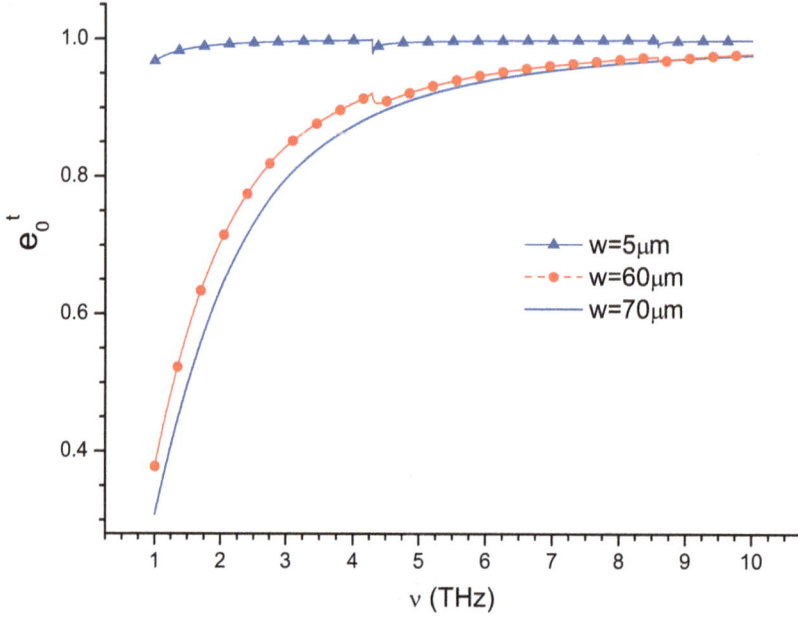

Figure 4.5. Efficiency e_0^t of the 0th transmitted–diffracted order (equation (4.9)) as a function of the frequency for free-standing graphene-strip gratings with $d = 70\ \mu m$ and strip widths $w = 5\ \mu m$ and $w = 65\ \mu m$. The transmissivity of the unpatterned graphene sheet ($w = d$) is given as a reference. The grating is normally illuminated by an s-polarized incident plane wave, $\varepsilon_1 = \varepsilon_2 = 1$, $\mu_1 = \mu_2 = 1$, $\mu_c = 0.39$ eV, $T = 300$ K and $\hbar\gamma = 0.658$ meV. The Python script used to obtain these curves is given in the appendix.

Figure 4.6 shows the spectral behavior of P_a and e_0^t for p polarization for a grating with strip width $w = 20\ \mu m$; the other parameters are equal to those considered in figures 4.4 and 4.5. For comparison, the curves for P_a and e_0^t, corresponding to the same grating illuminated under p polarization, and the curves corresponding to the absorption and transmittance of the infinite graphene sheet, are also shown. By contrast to the behavior exhibited for s polarization, and contrary to the expectation that a graphene sheet with void strips will have lower absorption and higher transmittance than an infinite planar graphene sheet without voids and with the same constitutive parameters, the p polarization curves in figure 4.6 exhibit enhanced absorption and almost null transmittance at a frequency $\nu \approx 2.57175$ THz. Other peaks of enhanced absorption and local minima in the efficiency of the 0th transmitted–diffracted order are observed at the frequency values $\nu \approx 5.2247, 6.92, 8.25$ and 9.41 THz. The Python script used to obtain the curves presented in figures 4.4, 4.5 and 4.6 can be found in the appendix.

4.4 Localized graphene surface plasmons

4.4.1 Resonance condition

An intuitive geometric picture for the peaks of enhanced absorption and local minima in the transmissivity observed in the p polarization curves in figure 4.6 can

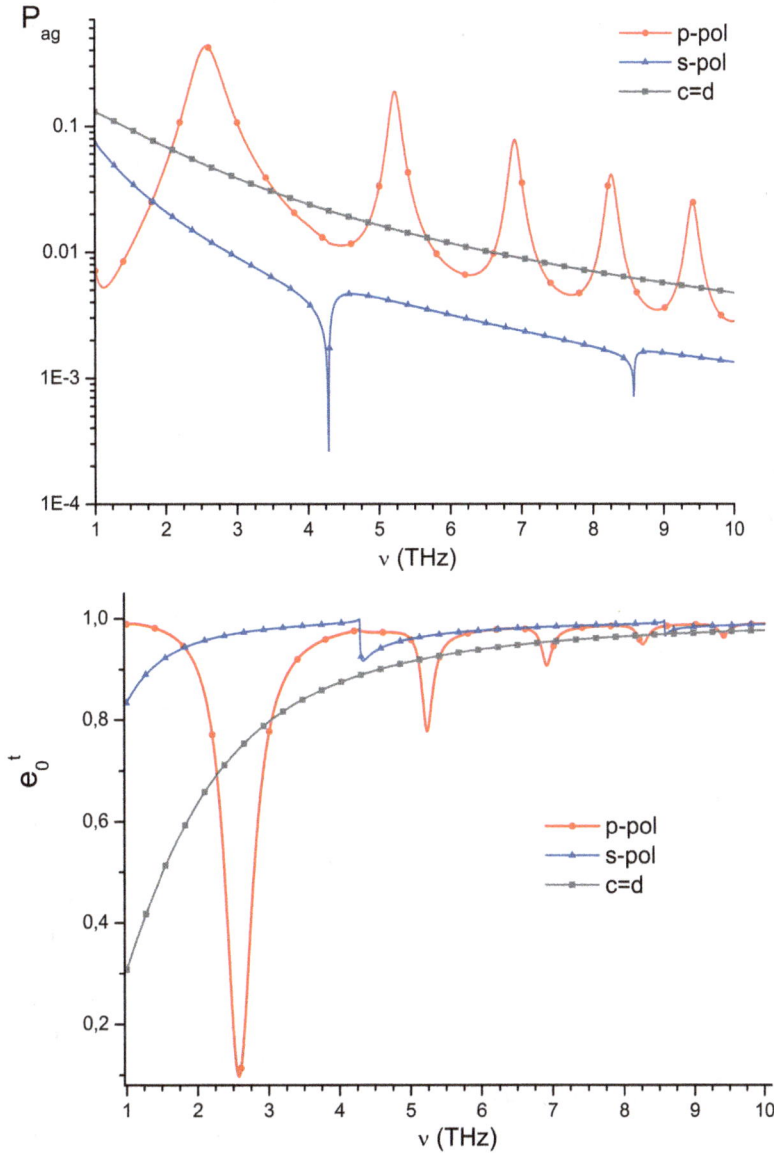

Figure 4.6. (a) Absorption P_a and (b) efficiency e_0^t of the 0th transmitted order as a function of the frequency for a free-standing graphene-strip grating normally illuminated by a linearly polarized incident plane wave. Red curves: s polarization; blue curves: p polarization; and gray curves: unpatterned graphene sheet. The values of the parameters are $d = 70$ μm, $w = 20$ μm, $\varepsilon_1 = \varepsilon_2 = 1$, $\mu_1 = \mu_2 = 1$, $\mu_c = 0.39$ eV, $T = 300$ K, $\hbar\gamma = 0.658$ meV, and correspond to the case considered in figure 11 of [6], figure 6 of [7] and figure 3 of [10]. The Python scripts used to obtain the curves in both polarization cases can be found in the appendix.

be obtained by recalling that an infinite self-standing graphene sheet immersed in a medium of relative constitutive parameters ε and μ may support propagating surface plasmons of the form given by equation (2.34). For p-polarized propagating surface plasmons to occur, Im $\tilde{\sigma} > 0$, a condition that in the collisionless zero temperature limit occurs when $0 < \hbar\omega/\mu_c < 1.667$. For the spectral range $0.011 < \hbar\omega/\mu_c < 0.11$ and for the values of the parameters considered in figure 4.6, Im $\bar{\sigma}$ is always positive and $\bar{\sigma}$ varies from a value of $\approx 4.66 + 29.28\,i$ for $\nu = 1$ THz to a value of $\approx 0.05 + 2.99\,i$ for $\nu = 10$ THz. In these conditions, if the graphene strips were infinitely wide ($w \to \infty$), they would support p-polarized propagating surface plasmons with a complex propagation constant κ given by equation (2.42). As the graphene strips are not infinitely wide, but have a finite size in the x direction, the strip edges act as discontinuities at which the propagating surface plasmon is partly reflected. As a result, a resonance is formed as a Fabry–Pérot-like standing wave at those frequencies which satisfy the condition [6, 10, 34]

$$\mathrm{Re}\,\kappa_{\mathrm{strip}}w = m\pi, \quad m = 1, 2, \ldots, \tag{4.30}$$

with κ_{strip} being the complex propagation constant of the finite strip. To avoid the calculation of κ_{strip} we assume that its value at the frequency ω does not differ appreciably from the value κ given by equation (2.42), which corresponds to an infinite graphene sheet. With this assumption, the following approximate resonance condition is obtained:

$$\mathrm{Re}\,\kappa w/\pi = 2w\,\mathrm{Re}\,\sqrt{\varepsilon\mu - 1/(2\pi\alpha\bar{\sigma}(\omega))^2}/\lambda \approx m, \quad m = 1, 2, \ldots, \tag{4.31}$$

with λ being the free-space wavelength.

The left-hand side of this condition, with $\bar{\sigma}(\omega)$ obtained from the Kubo model, is plotted as a function of frequency in figure 4.7 for various gratings with period $d = 70\,\mu\mathrm{m}$ and strip-widths of $w = 20, 30, 40, 50, 60$ and $65\,\mu\mathrm{m}$. The curve for $w = 20\,\mu\mathrm{m}$ corresponds to the grating considered in figure 4.6. For these quite narrow strips, the approximation involved in condition (4.4.2) is expected to hold in a qualitative manner. From the curves of P_a and e_0^t in figure 4.6 we observe that the approximation in fact overestimates the frequency of the first resonance, with $\mathrm{Re}\,\kappa w/\pi = 2w = 1$, predicting a value ≈ 3.07 THz, whereas the first peak of almost null transmittance and enhanced absorption occurs at ≈ 2.57175 THz. A comparison between the results obtained from the curve for $w = 20\,\mu\mathrm{m}$ in figure 4.7 (frequency values $\approx 5.48, 7.12, 8.44$ and 9.58 THz) and the spectral position of the resonances shown in figure 4.6 (frequency values $\approx 5.2247, 6.92, 8.30$ and 9.41 THz) shows that when condition (4.4.2) is applied to a grating with such narrow strips it also overestimates the spectral position of the higher order resonances with odd values of the index m. The overestimation becomes smaller for higher values of m (less than 2% for $m = 9$), which can be attributed to the fact that the wavelength of the higher order Fabry–Pérot-like standing waves becomes smaller and that it is this wavelength, and not the free-space wavelength, that serves as a characteristic length scale in establishing if the strips are wide enough for the approximation $\kappa_{\mathrm{strip}} \approx \kappa$ to hold.

Surface plasmon resonances with even values of the integer m in equation (4.4.2) have an odd field symmetry with respect to the strip center [34]. Therefore, resonance frequencies obtained from even values in the vertical axis of figure 4.7 are dark modes at normal incidence, which is why resonances with even values of m do not appear in the curves of P_a and e_0^t in figure 4.6. However, surface plasmon modes

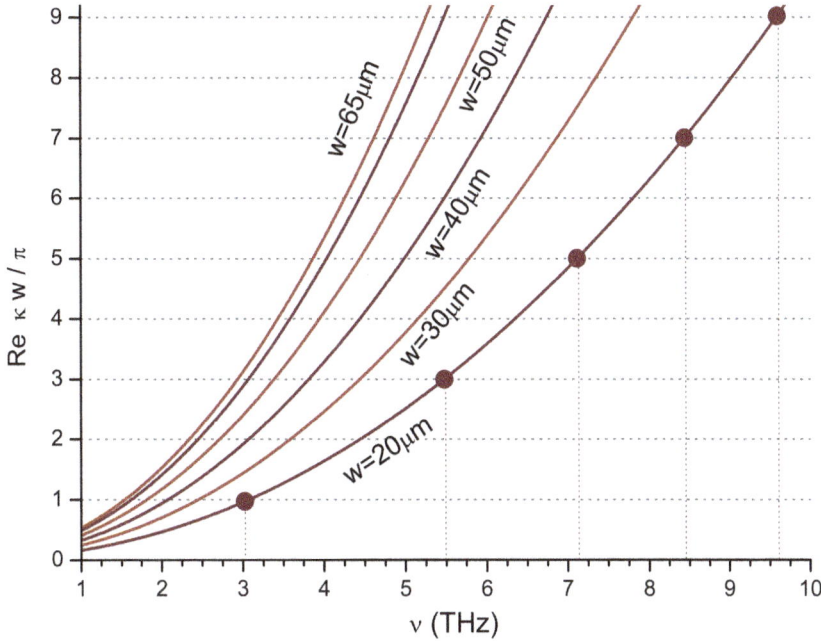

Figure 4.7. Re $\kappa w/\pi = 2w$, the left-hand side of the surface plasmon resonance condition (4.4.2), as a function of frequency for various free-standing graphene-strip gratings with the same period $d = 70$ μm and strip-widths of $w = 20, 30, 40, 50, 60$ and 65 μm. The other parameters are as in figure 4.6, where surface plasmon resonances for a grating with $w = 20$ μm can be observed in normal incidence at $\nu \approx 2.57175, 5.2247, 6.92, 8.30$ and 9.41 THz.

with odd field symmetry with respect to the strip center can be excited in oblique incidence, as shown in figure 4.8, obtained for the same parameters considered in figure 4.6, except for the angle of incidence, which is now $\theta_0 = 30°$. Apart from the resonances corresponding to odd integer values in the vertical axis of the curve for $w = 20$ μm in figure 4.7, new resonances now appear in the curves for P_a and e_0^t in figure 4.8. The lower resonances not present for normal incidence are located near $\approx 4.12, 6.10, 7.48$ and 8.30 THz. They correspond to surface plasmon modes with odd field symmetry with respect to the center of the strip, that is, to solutions with even values of the index m in the resonance condition (4.4.2). For $m = 2, 4, 6$ and 8, figure 4.7 gives frequency values near $\approx 4.44, 6.35, 7.81$ and 9.02 THz, respectively, which agree quite well with the values obtained from the position of the resonances, despite the relatively low occupation factor of the strips ($w/d < 28\%$).

The slopes of the curves in figure 4.7 indicate that the higher the value of the the strip grating's occupation factor w/d, the greater the number of graphene surface plasmon resonances in a given frequency interval. This can be appreciated in figure 4.9, where the curves of P_{ag} and e_0^t are plotted as functions of the frequency in the interval between 1 THz and 10 THz for three gratings with identical constitutive and geometrical parameters, except for the widths of the strips, which are $c = 30$ μm (top, with eight resonances), 50 μm (middle, with 12 resonances) and 65 μm (bottom,

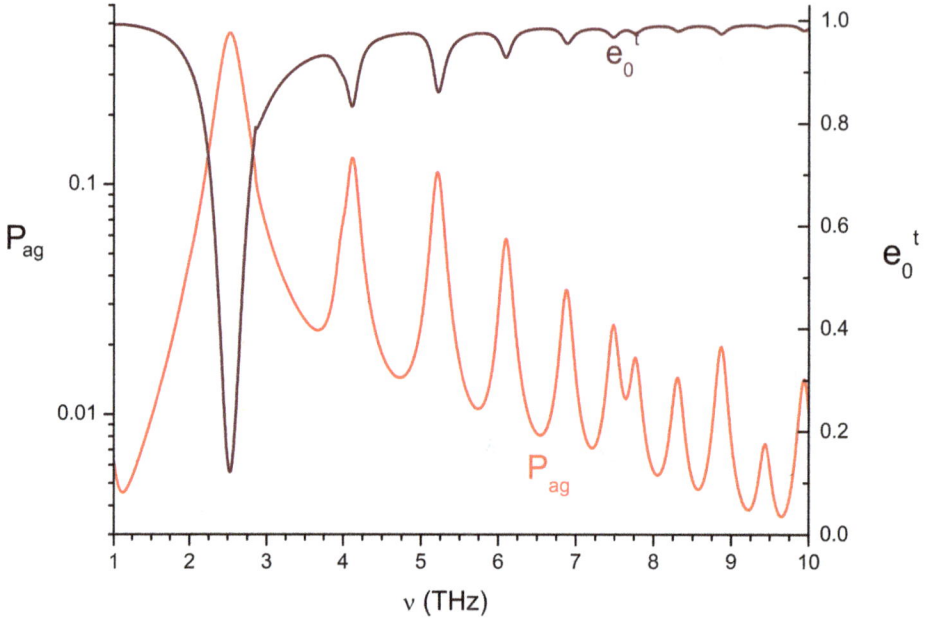

Figure 4.8. Absorption P_a and efficiency e_0^t in p polarization of the 0th transmitted order for a grating with the same parameters as those in figure 4.6, except that $\theta_0 = 30°$ (oblique incidence). Apart from the resonance peaks observed for normal incidence, new resonance peaks, corresponding to even values in the vertical axis of figure 4.7, appear. The Python script used to obtain these curves can be found in the appendix.

15 resonances). Due to the fact that the illumination is normal to the plane that contains the strips, only the surface plasmon modes with odd values of the index m can be excited. Incidentally, observe that compared with the flat graphene sheet (gray curves in figure 4.9), the absorption P_{ag} is always appreciably enhanced for frequencies after the first resonance.

4.4.2 Near field

The near-field patterns for the magnetic field diffracted from a strip grating (strip width $w = 20$ μm, period $d = 70$ μm) illuminated by p-polarized incident radiation are shown in figure 4.10. The frequency values correspond to the first four enhanced absorption peaks that occur in figure 4.8. Taking into account the fact that these numerical results were obtained after truncation of the system of linear equations (4.28), and that this system of equations was inferred without the explicit invocation of any standing-wave mechanism, the near field patterns shown in figure 4.10 are a clear indication that the occurrence of enhanced absorption in this grating when the frequency of the incident radiation is given approximately by condition (4.31) is a phenomenon that is intimately linked with the excitation of standing waves in the grating strips.

For the frequency value $\nu = 2.57$ Thz, which corresponds to the value $m = 1$ in the vertical axis of figure 4.7, figure 4.10 shows that the spatial distribution for the

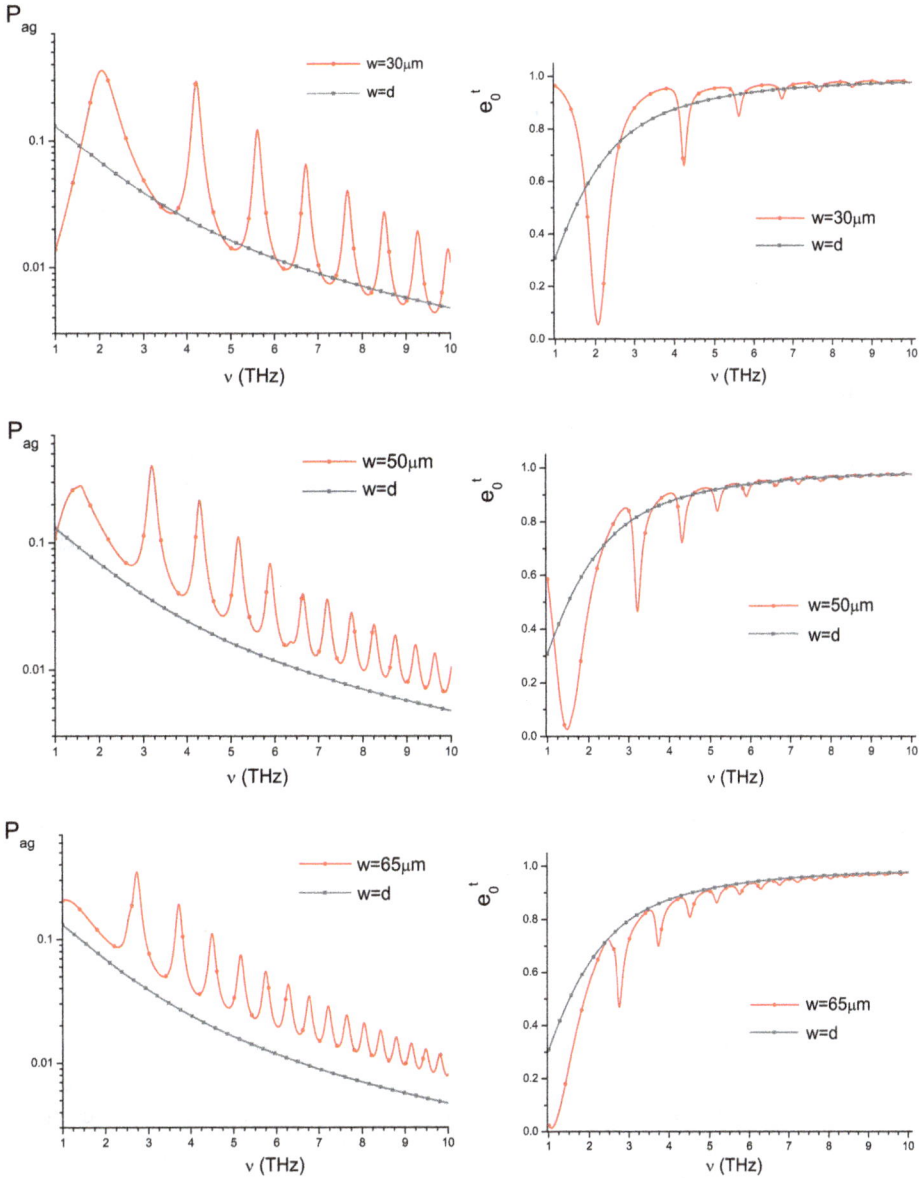

Figure 4.9. Curves of P_{ag} (left column) and e_0^t (right column) vs frequency for three different gratings with $c = 30$ μm (top), $c = 50$ μm (center) and $c = 65$ μm (bottom) illuminated under normal incidence in p polarization. The other parameters are the same as those in figure 4.6. The gray curves correspond to an infinite graphene sheet.

absolute value of the magnetic field $|f(x, 0)|$ along the strip ($0 \leqslant x \leqslant 20$ μm) does not have any node points, as expected for the fundamental oscillation mode of 1D vibrating systems. Analogously, for the frequency values $\nu = 4.12$, 5.23 and 6.10 Thz, which correspond to the values $m = 2$, 3 and 4 in the vertical axis of figure 4.7,

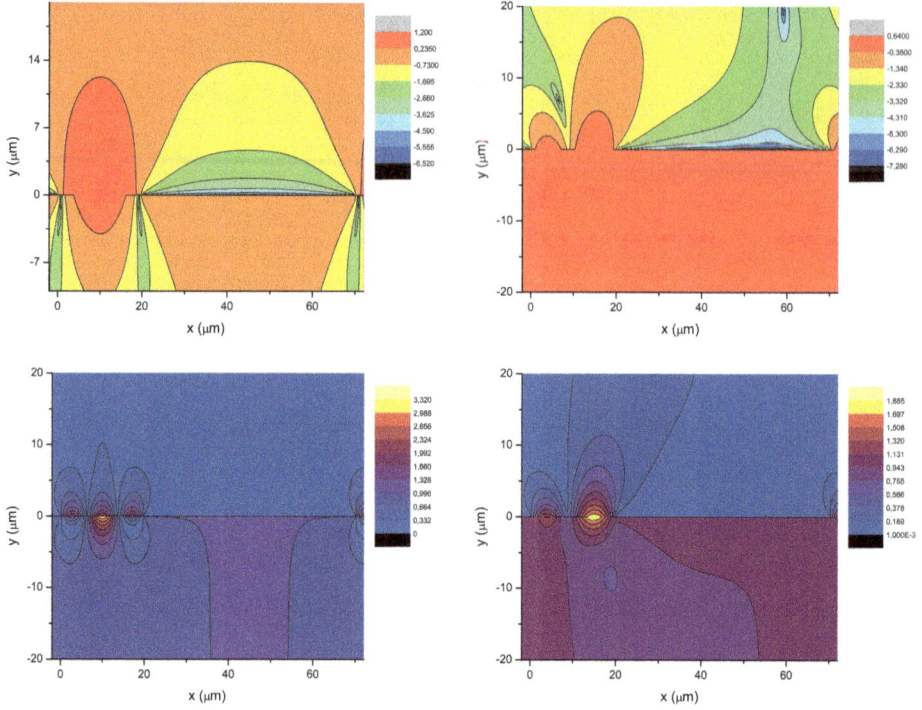

Figure 4.10. Near-field patterns for the diffracted magnetic field (p polarization) for a grating, strip width $w = 20$ μm, period $d = 70$ μm. The frequency of the incident radiation corresponds to $\nu = 2.57$, 4.12, 5.23 and 6.10 Thz, that is to the first four resonances shown in figure 4.8. The color map indicates the magnitude of $|f(x, y)|$. Left column: normal incidence. Right column: $\theta_0 = 30°$. The other parameters are the same as those in figure 4.6.

respectively, the spatial distributions for the absolute value of the magnetic field $|f(x, 0)|$ along the strip have $m - 1$ nodal points, as expected for the mth oscillation mode of 1D vibrating systems and in complete agreement with the geometric model relying on the existence of Fabry–Pérot-like standing waves.

The *propagating* surface plasmons corresponding to a flat graphene sheet discussed in section 2.3 play a fundamental role in the geometric model used here to explain simply and clearly the *localized* surface plasmon effects produced in the diffraction response of graphene strip gratings. This kind of intuitive geometrical model is not confined to localized surface plasmons in graphene strips and it can actually be applied to localized surface plasmons in other graphene structures, such as graphene-coated subwavelength wires [35], or even to other localized surface plasmons, such as metallic nanowires [36] or coaxial metallic structures [37].

4.4.3 Numerical aspects

In principle, the linear systems (4.23) and (4.28) can be solved numerically after truncation. The number NMAX of terms retained in the diffracted field expansions should be established to guarantee convergence of the results (see, e.g. [1, 7, 10, 12]). However, the main disadvantage of most Fourier-based methods is that they exhibit numerical problems in efficiently describing sharp variations such as those that occur at the edge of the strips. As a consequence, convergence is achieved with rather large matrices [7, 10].

For perfectly conducting strips, the slow convergence of the modal method has been improved by using the concept of adaptive spatial resolution, which consists of introducing a new coordinate system in which the fields are meshed more finely in the neighborhood of the strips. For both flat [38] and curved perfectly conducting strips [25], the use of adaptive spatial resolution has resulted in a better representation of the fields, with a dramatic reduction of the Gibbs phenomenon. The adaptive spatial resolution technique provides a route to improve the convergence of the modal method for graphene strip gratings. Another approach, based on the analytical regularization technique [10], leads to a numerical scheme where the convergence of the results after truncation of the modal field expansions to finite order is guaranteed in a mathematical sense, that is, solutions obtained with progressively larger orders of truncation lead to smaller errors. These approachs will not be discussed here, the interested reader is referred to [10, 25].

For the results presented here, typical values for the truncation parameter are NMAX = 31 (that is, with space harmonics indexes $|n| < 15$) for s polarization (equation (4.23)) and NMAX = 501 (space harmonics indexes $|n| < 250$) for p polarization (equation (4.28)). For these values the results converge to within 1% (s polarization) and 5% (s polarization).

The constitutive and geometrical parameters used in figures 4.6 and 4.8 are the same as those used in figure 11 of [6] to illustrate the performance of a singular integral equation method. These results were later taken as a benchmark for checking the accuracy of the numerical methods used in [7] (figure 6) and [10] (figure 3). Very good agreement is observed between the curves shown in figures 4.6 and 4.8 for both polarization cases—obtained from the straightforward implementation of the mode-matching method by Fourier expansion—and those obtained in [6, 7, 10] using more sophisticated numerical calculations. However, the reader should be warned that the straightforward implementation that is used here to show one of the simplest mathematical formulations for graphene gratings does not always give accuracy controlled by the size of the matrix equation and does not guarantee machine precision convergence, as more sophisticated methods do.

4.5 Propagating graphene surface plasmons

We have seen that a graphene sheet with void strips supports localized surface plasmons, which, according to the approximate resonance condition (4.4.2) can be modeled as standing waves localized at the graphene strips. If the void strips are replaced by graphene strips with a different surface conductivity value, a structure of allowed and forbidden bands appears for the surface plasmons that can propagate along successive strips. Although the mode-matching technique is not adequate for studying the rigorous band structure of surface plasmons in systems composed of periodic patches of graphene, the technique can be used when the flat grating has continuous variations of conductivity [1]. Although the solution of the homogeneous problem for conductivity gratings is outside the scope of this concise introduction, we can take advantage of the mode-matching solution to show that these propagating graphene surface plasmons can be resonantly excited in diffraction experiments.

Let us consider a flat graphene sheet with a periodically modulated optical conductivity characterized by $u(x) = 1 + h \cos Gx$, with $G = 2\pi/d$ and h a parameter denoting the strength of the modulation. In this case, the coefficients $L(p)$ in equation (4.22) are given analytically as

$$L(p) = \delta_{p,0} + \frac{h}{2}(\delta_{p,1} + \delta_{p,-1}). \tag{4.32}$$

As the value of the position-dependent conductivity is given by $\sigma u(x)$, for very low values of h we expect the conductivity grating to support a graphene surface plasmon with a propagation constant that is very close to the value $\kappa(h = 0)$ corresponding to the uniformly doped graphene sheet. As discussed in section 3.1, the projection along the interface of the wave vector of any incident plane wave is always smaller than $\kappa(h = 0)$, even at a grazing incidence where the projection takes its maximum value. However, when $h \neq 0$, the modulation produces a coupling between the space harmonics and therefore phase matching between the incident wave and SPPs can occur if there is a value α_n equal to the real part of $\kappa(h \neq 0) \approx \kappa(h = 0)$. This is the working principle of the grating coupler, a convenient alternative to ATR couplers, particularly for p-polarized graphene SPPs that, as shown in figure 3.1, may have κ values that are too large to be excited using prism couplers.

As an example, we consider that the sinusoidally modulated conductivity grating is immersed in air and has a period $d = 2.32$ µm, an average dimensionless conductivity $\tilde{\sigma} = 0.00647601 + 3.08871i$, a value obtained from the Kubo model for an incident frequency $\nu = 19.88$ THz (vacuum wavelength $\lambda = 15.09$ µm), an absolute temperature $T = 300$K, a collision frequency $\gamma = 0.1$ meV and a chemical potential $\mu_c = 0.8$ eV. In this case, equation (2.42) predicts the value $\kappa(h = 0) = (7.06114 + 0.0148049i) \, \omega/c$. According to the previous discussion, the approximated coupling condition gives $\alpha_n \approx 7.06114 \, \omega/c$ and using equation (4.3) we obtain that for these grating parameters p-polarized SPPs may be resonantly excited by the incident wave when the angle of incidence is $\theta_0 \approx 34.13°$. In general, the approximated coupling condition for a graphene sheet immersed in a medium of relative constitutive parameters ε and μ can be written as

$$\sin \theta_0 \approx \pm n\frac{\lambda}{d} + \mathrm{Re} \sqrt{\varepsilon\mu - 1/(2\pi\alpha\tilde{\sigma}(\omega))^2}, \quad n = 1, 2 \ldots \tag{4.33}$$

To check these predictions, in figure 4.11 we show the graphene absorption (P_{ag}) and efficiency of the 0th transmitted order (e_0^t) as functions of the angle of incidence θ_0

for flat graphene sheets with $\tilde{\sigma} = 0.00647601 + 3.08871i$, $d = 2.32$ μm, $\nu = 19.88$ Thz ($\lambda/d = 6.5$) and $h = 0.05, 0.1, 0.2$ and 0.3. The Python script used to obtain these curves can be found in the appendix. The p-polarized transmissivity of the flat graphene sheet with uniform ($h = 0$) surface conductivity is also given as a reference. For the lowest value of h considered, a minimum in e_0^t and a maximum in P_{ag} occur at an angle of incidence $\theta_0 \approx 39.8°$, close but not equal to the value predicted by equation (4.33). The similarity between the observed and predicted angular positions indicates the degree of adequacy of the assumption used to derive equation (4.33), namely, that the propagating surface plasmon of the modulated conductivity sheet has a propagation constant $\kappa(h)$ that is equal to the value $\kappa(0)$ corresponding to a uniformly doped graphene sheet with a surface conductivity equal to the average modulation. A minimum in e_0^t and a maximum in P_{ag} can also be observed at the same angular position when the amplitude of the modulation increases and all other parameters remain constant. However, the discrepancies between the observed and predicted angular positions become larger for stronger modulations.

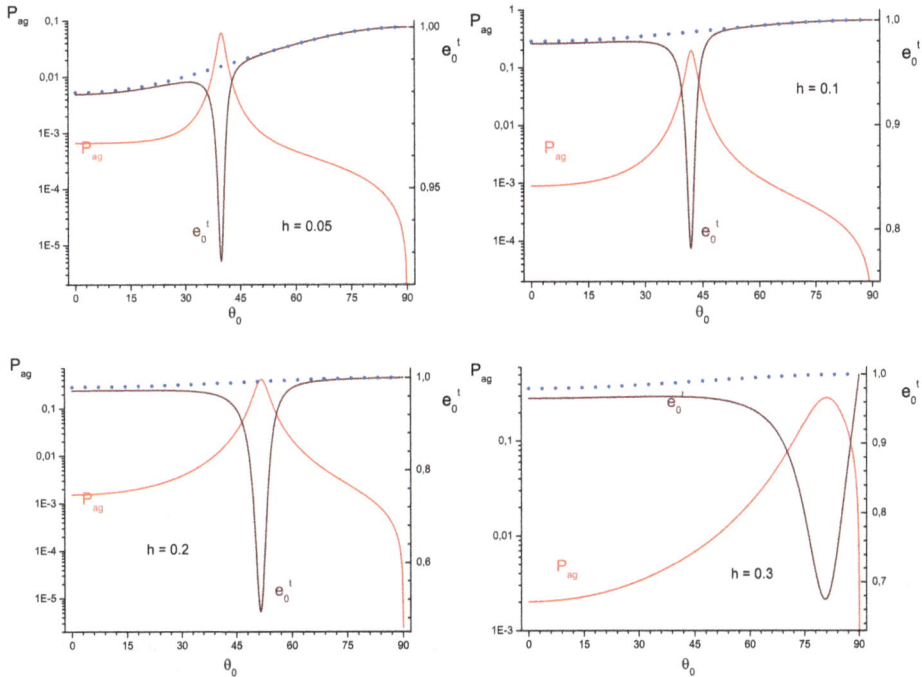

Figure 4.11. Graphene absorption (P_{ag}) and efficiency of the 0th transmitted order (e_0^t) as functions of the angle of incidence θ_0 for conductivity gratings with $u(x) = 1 + h \cos Gx$, $G = 2\pi/d$, $d = 2.32$ μm and four values of the modulation parameter, $h = 0.05, 0.1, 0.2$ and 0.3. The incident wave is p-polarized and the average dimensionless conductivity is $\tilde{\sigma} = 0.00647601 + 3.08871i$, a value corresponding to Kubo parameters $\nu = 19.88$ Thz (vacuum wavelength $\lambda = 15.09$ μm), $T = 300$ K, $\gamma = 0.1$ meV and $\mu_c = 0.8$ eV. The dots indicate the p-polarized transmissivity of the flat graphene sheet with uniform ($h = 0$) surface conductivity. The Python script used to obtain these curves can be found in the appendix.

References

[1] Bludov Y V, Ferreira A, Peres N M R and Vasilevskiy M I 2013 A primer on surface plasmon-polaritons in graphene *Int. J. Mod. Phys.* B **27** 1341001

[2] Fei Z *et al* 2011 Infrared nanoscopy of dirac plasmons at the Graphene-SiO$_2$ interface *Nano Lett.* **11** 4701–5

[3] Filter R, Farhat M, Steglich M, Alaee R, Rockstuhl C and Lederer F 2013 Tunable graphene antennas for selective enhancement of THz emission *Opt. Express* **21** 3737–45

[4] Vasic B, Isic G and Gajic R 2013 Localized surface plasmon resonances in graphene ribbon arrays for sensing of dielectric environment at infrared frequencies *J. Appl. Phys.* **113** 113110

[5] Li K, Ma X, Zhang Z, Xu Y and Song G 2014 Tunable and angle-insensitive plasmon resonances in graphene ribbon arrays with multispectral diffraction response *J. Appl. Phys.* **115** 104306

[6] Shapoval O V, Gomez-Diaz J S, Perruisseau-Carrier J, Mosig J R and Nosich A I 2013 Integral equation analysis of plane wave scattering by coplanar graphene-strip gratings in the THz range *IEEE Trans. Terahertz Sci. Technol.* **3** 666–74

[7] Hwang R-B 2014 Rigorous formulation of the scattering of plane waves by 2D graphene-based gratings: out-of-plane incidence *IEEE Trans. Antennas Propag.* **62** 4736–45

[8] He X and Lu H 2014 Graphene-supported tunable extraordinary transmission *Nanotechnology* **25** 325201

[9] He X 2015 Tunable terahertz graphene metamaterials *Carbon* **82** 229–37

[10] Zinenko T I 2015 Scattering and absorption of terahertz waves by a free-standing infinite grating of graphene strips: analytical regularization analysis *J. Opt.* **17** 055604

[11] Peres N M R, Ferreira A, Bludov Y V and Vasilevskiy M I 2012 Light scattering by a medium with a spatially modulated optical conductivity: the case of graphene *J. Phys.: Condens. Matter* **24** 245303

[12] Slipchenko T M, Nesterov M L, Martin-Moreno L and Nikitin A Y 2013 Analytical solution for the diffraction of an electromagnetic wave by a graphene grating *J. Opt.* **15** 114008

[13] Davoyan A R, Popov V V and Nikitov S A 2012 Tailoring terahertz near-field enhancement via two-dimensional plasmons *Phys. Rev. Lett.* **108** 127401

[14] Xiao S and Mortensen N A 2011 Surface-plasmon-polariton-induced suppressed transmission through ultrathin metal disk arrays *Opt. Lett.* **36** 37

[15] Peres N M R, Bludov Y V, Ferreira A and Vasilevskiy M I 2013 Exact solution for square-wave grating covered with graphene: surface plasmon-polaritons in the terahertz range *J. Phys.: Condens. Matter* **25** 125303

[16] Ju L *et al* 2011 Graphene plasmonics for tunable terahertz metamaterials *Nature Nanotech.* **6** 630–4

[17] Nikitin A Y, Guinea F, Garcia-Vidal J and Martin-Moreno L 2012 Surface plasmon enhanced absorption and suppressed transmission in periodic arrays of graphene ribbons *Phys. Rev.* B **85** 081405(R)

[18] Yan H, Li X, Chandra B, Tulevski G, Wu Y, Freitag M, Zhu W, Avouris P and Xia F 2012 Tunable infrared plasmonic devices using graphene/insulator stacks *Nature Nanotech.* **7** 330–4

[19] Thongrattanasiri S, Koppens F H L and García De Abajo J 2012 Complete optical absorption in periodically patterned graphene *Phys. Rev. Lett.* **108** 047401

[20] Zhan T R, Zhao F Y, Hu X H, Liu X H and Zi J 2012 Band structure of plasmons and optical absorption enhancement in graphene on subwavelength dielectric gratings at infrared frequencies *Phys. Rev.* B **86** 165416

[21] Waterman P C 1975 Scattering by periodic surfaces *J. Acoust. Soc. Am.* **57** 791–802
[22] Maystre D (ed) 1992 *Selected Papers on Diffraction Gratings* (Bellingham, WA: SPIE)
[23] Petit R (ed) 1980 *Electromagnetic Theory of Gratings* (New York: Springer)
[24] Loewen E and Popov E E 1997 *Diffraction Gratings and Applications* (New York: Dekker)
[25] Sirenko Y K and Ström S (ed) 2010 *Modern Theory of Gratings. Resonant Scattering: Analysis Techniques and Phenomena* (New York: Springer)
[26] Depine R A and Lakhtakia A 2004 Plane-wave diffraction at the periodically corrugated boundary of vacuum and a negative-phase-velocity material *Phys. Rev.* E **69** 057602
[27] Depine R A and Lakhtakia A 2004 Perturbative approach for diffraction due to a periodically corrugated boundary between vacuum and a negative phase-velocity material *Opt. Commun.* **233** 277–82
[28] Depine R A and Lakhtakia A 2005 Diffraction gratings of isotropic negative phase-velocity materials *Optik* **116** 31–43
[29] Rayleigh L (J. W. Strutt) 1907 On the dynamical theory of gratings *Proc. R. Soc. London Ser.* A **79** 399–416
[30] Tishchenko A V 2010 Rayleigh was right: Electromagnetic fields and corrugated interfaces *Opt. Photon. News* **21** 50–4
[31] Tishchenko A V 2009 Numerical demonstration of the validity of the Rayleigh hypothesis *Opt. Express* **17** 17102–17
[32] Depine R A and Gigli M L 1994 Diffraction from corrugated gratings made with uniaxial crystals: Rayleigh methods *J. Mod. Opt.* **41** 695–715
[33] Zinenko T L, Nosich A I and Okuno Y 1998 Plane wave scattering and absorption by resistive-strip and dielectric strip periodic gratings *IEEE Trans. Antennas Propagat.* **46** 1498–505
[34] Balaban M V, Shapoval O V and Nosich A I 2013 THz wave scattering by a graphene strip and a disk in the free space: integral equation analysis and surface plasmon resonances *J. Opt.* **15** 114007
[35] Cuevas M, Riso M A and Depine R A 2016 Complex frequencies and field distributions of localized surface plasmon modes in graphene-coated subwavelength wires *J. Quant Spectrosc. Radiat* **114** 26–33
[36] Schmidt M A and Russell P S J 2008 Long-range spiralling surface plasmon modes on metallic nanowires *Opt. Express* **16** 13617
[37] Catrysse P B and Fan S H 2009 Understanding the dispersion of coaxial plasmonic structures through a connection with the planar metal-insulator-metal geometry *Appl. Phys. Lett.* **94** 231111
[38] Granet G and Guizal B 2005 Analysis of strip gratings using a parametric modal method by Fourier expansions *Opt. Commun.* **255** 1–11

Chapter 5

Graphene wires

In previous chapters we focused on the electromagnetic description of graphene structures associated with different degrees of symmetry, starting from the flat and homogeneously doped sheets of chapters 2 and 3, which exhibit two-dimensional (2D) continuous translational symmetry, then moving on to the one-dimensional (1D) photonic crystals of section 3.4, which exhibit 2D continuous translational symmetry along the graphene sheets and 1D discrete translational symmetry along the normal direction, before considering the gratings of chapter 4, which exhibit 1D continuous translational symmetry along one tangent direction on the graphene sheet and 1D discrete translational symmetry along the other. In this chapter we consider the interaction between electromagnetic radiation and graphene wires, that is structures which keep the 1D continuous translational symmetry along the wire axis, but do not have any translational symmetry in the plane perpendicular to this axis.

5.1 2D graphene particles

Recent experiments [1–4] have shown that a graphene sheet can be coated uniformly and tightly on the surface of a material particle via the van der Waals force. A graphene-coated wire can be used to propagate radiation along its axis; as a waveguide; or for sensing and laser applications [5]. Apart from their particular waveguiding characteristics [6–11], graphene wires can be used to increase the interaction length and counter the limited absorption in a graphene monolayer, a problem in some applications, for instance in optical modulators [12]. Instead of looking at applications where the electromagnetic radiation propagates along the wire axis, in this chapter we consider that the wire is illuminated perpendicular to its axis. This case provides a convenient 2D model for illustrating the light scattering properties of graphene particles without the complications caused by polarization

doi:10.1088/978-1-6817-4309-7ch5

vectors in other scattering geometries. This is because the problem considered here shares the 1D continuous translational symmetry of all the problems discussed in this book, and therefore the vectorial electromagnetic scattering problem can be handled in a scalar way, that is, the solution to any incident polarization can be described as a linear combination of the familiar scalar problems for s polarization (electric field parallel to wire axis) and p polarization (magnetic field parallel to wire axis). The truly vectorial problem of scattering from a spherical object coated by a graphene layer is treated in [13].

It is well known that metallic particles support localized surface plasmons, that is, localized resonances characterized by a set of discrete resonant frequencies which, under adequate circumstances, can be coupled with incident radiation without the necessity of additional coupling structures [14–19]. With graphene being a plasmonic material, we expect graphene-coated particles to support localized surface plasmons too. In fact, we already found a very simple particular case of localized surface plasmons on flattened particles in section 4.4, when we were considering the electromagnetic response of a periodic arrangement of free-standing graphene ribbons. It should be noted that a graphene coating on a nonplasmonic particle is expected to introduce localized surface plasmon mechanisms that do not exist in the bare particle. Analogously, a graphene coating on a metallic or metal-like particle is expected to modify the localized surface plasmons that already exist in the bare particle. In both cases, the good tunability of the graphene coating leads to unprecedented control over the location and magnitude of the particle plasmonic resonances [13, 20, 21]. In this chapter we give a rigorous electromagnetic analysis of the scattering of a linearly polarized plane wave by circular cross-section wires coated with a graphene monolayer, a system that is not only interesting for several applications, but also provides a simple canonical model for the scattering character-istics of other graphene-coated particles.

5.2 Scattering problem

We consider a graphene-coated cylinder with a circular cross-section (radius R) centered at $x = 0$, $y = 0$ (see figure 5.1). The wire substrate may be dielectric or conducting (electric permittivity ε_1 and magnetic permeability μ_1), and the coated wire is embedded in a transparent medium (electric permittivity ε_2 and magnetic

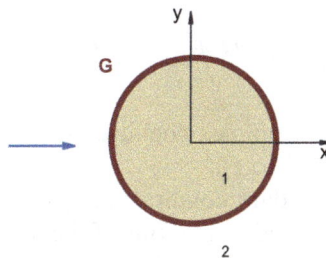

Figure 5.1. A graphene-coated wire with a circular cross-section is illuminated perpendicular to its axis.

permeability μ_2). The propagation of the linearly polarized incident plane wave is directed along z. As in the problems considered in previous chapters, any incident polarization can be described as a linear combination of the solutions obtained in two fundamental scalar problems: s polarization, when the electric field of the incident field is directed along the wire axis (z), and p polarization, when the magnetic field of the incident field is directed along the wire axis. In the first case the magnetic field remains parallel to the main section of the cylindrical surface, and in the second case it is the electric field that remains parallel to this section. We continue to denote the non-zero component of the total electromagnetic field along the z direction for each polarization case as f.

To obtain analytical solutions to the scattering problem, the usual separation-of-variables approach for graphene-free cylinders [23, 24] can be followed very closely, except for the boundary conditions, which now include new terms that are proportional to the surface conductivity σ. In the first step, and after changing from the Cartesian coordinates $x-y$ to the usual polar coordinates $\rho-\varphi$, we expand $f(\rho, \varphi)$ as a series of cylindrical harmonics, one for the internal region ($\rho < R$, subscript 1) and another one for the external region ($\rho > R$, subscript 2). When the incident magnetic field is contained in the $x-y$ plane (s polarization), the expansions for the total electric field along the axis of the cylinder are written as

$$f_1(\rho, \varphi) = \sum_{n=-\infty}^{\infty} d_n J_n(k_1\rho) \exp in\varphi, \tag{5.1}$$

$$f_2(\rho, \varphi) = \sum_{n=-\infty}^{\infty} \left[B_0 i^n J_n(k_2\rho) + b_n H_n^{(1)}(k_2\rho) \right] \exp in\varphi, \tag{5.2}$$

where b_n and d_n are unknown complex coefficients, $k_j = \frac{\omega}{c}\sqrt{\varepsilon_j\mu_j}$ ($j = 1, 2$), c is the speed of light in vacuum, B_0 is the amplitude of the incident electric field (parallel to z), and J_n and $H_n^{(1)}$ are the nth Bessel and Hankel functions of the first kind, respectively. In the same manner, when the incident electric field is contained in the $x-y$ plane (p polarization), the expansions for the total magnetic field along the axis of the cylinder are

$$f_1(\rho, \varphi) = \sum_{n=-\infty}^{\infty} c_n J_n(k_1\rho) \exp in\varphi, \tag{5.3}$$

$$f_2(\rho, \varphi) = \sum_{n=-\infty}^{\infty} \left[A_0 i^n J_n(k_2\rho) + a_n H_n^{(1)}(k_2\rho) \right] \exp in\varphi, \tag{5.4}$$

where a_n and c_n are unknown complex coefficients, and A_0 is the amplitude of the incident magnetic field (parallel to z).

In the second step, the boundary conditions at $\rho = R$ are invoked to obtain the unknown coefficients a_n, b_n, c_n and d_n in terms of the incident amplitudes A_0 and B_0. Because of the 1D continuous translational symmetry along the z direction that this problem and the problem considered in section 4.1.2 share, in both cases the boundary conditions take the form given by equations (4.6) and (4.7), except that (i) for the grating problem the curve is given by $y = g(x)$, whereas for the graphene-coated cylinder the curve is the circle $\rho = R$; (ii) here the function u in section 4.1.2 must be set equal to unity, since the conductivity of the coating is assumed to be constant along $\rho = R$; and (iii) the unit vector normal to the circle and pointing from medium 2 to medium 1 is $n = -\rho$. When these differences are taken into account, the boundary conditions at $\rho = R$ can be written as

$$\left.\begin{array}{l} f_1(R, \varphi) = f_2(R, \varphi) \\[2ex] \dfrac{1}{\mu_2}\dfrac{\partial f_2}{\partial \rho}(R, \varphi) - \dfrac{1}{\mu_1}\dfrac{\partial f_1}{\partial \rho}(R, \varphi) = -i\omega\mu_0\sigma f_1(R, \varphi) \end{array}\right\} \quad s \text{ polarization} \qquad (5.5)$$

and

$$\left.\begin{array}{l} \dfrac{1}{\varepsilon_1}\dfrac{\partial f_1}{\partial \rho}(R, \varphi) = \dfrac{1}{\varepsilon_2}\dfrac{\partial f_2}{\partial \rho}(R, \varphi) \\[2ex] f_2(R, \varphi) - f_1(R, \varphi) = i\dfrac{\sigma}{\omega\varepsilon_0}\dfrac{1}{\varepsilon_1}\dfrac{\partial f_1}{\partial \rho}(R, \varphi) \end{array}\right\} \quad p \text{ polarization} \qquad (5.6)$$

As in the case of graphene-free cylinders [23, 24], the expansions for the fields are introduced into the boundary conditions at $\rho = R$, and, using the orthogonality of the angular harmonics $\exp in\varphi$, $-\infty < n < \infty$ in the interval $0 < \varphi < 2\pi$, the amplitudes of the scattered field can be written as

$$a_n = \frac{-i^n\left[\dfrac{\varepsilon_1}{k_1}J_n(x_1)J_n'(x_2) - \dfrac{\varepsilon_2}{k_2}J_n'(x_1)J_n(x_2) + 4\pi\alpha\,\tilde{\sigma}\dfrac{c}{\omega}iJ_n'(x_1)J_n'(x_2)\right]}{\dfrac{\varepsilon_1}{k_1}J_n(x_1)H_n'^{(1)}(x_2) - \dfrac{\varepsilon_2}{k_2}J_n'(x_1)H_n^{(1)}(x_2) + 4\pi\alpha\,\tilde{\sigma}\dfrac{c}{\omega}iJ_n'(x_1)H_n'^{(1)}(x_2)}A_0 \qquad (5.7)$$

$$b_n = \frac{-i^n\left[\dfrac{\varepsilon_2}{k_2}J_n(x_1)J_n'(x_2) - \dfrac{\varepsilon_1}{k_1}J_n'(x_1)J_n(x_2) + 4\pi\alpha\,\tilde{\sigma}\dfrac{c}{\omega}iJ_n(x_1)J_n(x_2)\right]}{\dfrac{\varepsilon_2}{k_2}J_n(x_1)H_n'^{(1)}(x_2) - \dfrac{\varepsilon_1}{k_1}J_n'(x_1)H_n^{(1)}(x_2) + 4\pi\alpha\,\tilde{\sigma}\dfrac{c}{\omega}iJ_n(x_1)H_n^{(1)}(x_2)}B_0, \qquad (5.8)$$

while the amplitudes of the fields inside the wire are given by [20, 21]

$$c_n = \frac{\dfrac{\varepsilon_1}{k_1}i^n\left[J_n(x_2)H_n'^{(1)}(x_2) - J_n'(x_2)H_n^{(1)}(x_2)\right]}{\dfrac{\varepsilon_1}{k_1}J_n(x_1)H_n'^{(1)}(x_2) - \dfrac{\varepsilon_2}{k_2}J_n'(x_1)H_n^{(1)}(x_2) + 4\pi\alpha\,\tilde{\sigma}\dfrac{c}{\omega}iJ_n'(x_1)H_n'^{(1)}(x_2)}A_0 \qquad (5.9)$$

$$d_n = \frac{\dfrac{\varepsilon_2}{k_2}i^n\left[J_n(x_2)H_n'^{(1)}(x_2) - J_n'(x_2)H_n^{(1)}(x_2)\right]}{\dfrac{\varepsilon_2}{k_2}J_n(x_1)H_n'^{(1)}(x_2) - \dfrac{\varepsilon_1}{k_1}J_n'(x_1)H_n^{(1)}(x_2) + 4\pi\alpha\,\tilde{\sigma}\dfrac{c}{\omega}iJ_n(x_1)H_n^{(1)}(x_2)}B_0, \qquad (5.10)$$

where $x_1 = k_1R$, $x_2 = k_2R$. It can be checked that the factor between the square brackets in the numerator of the coefficients c_n and d_n can be rewritten in terms of the Wronskian $W(J_n, H_n^{(1)})(x_2)$, and that the multipole coefficients have essentially the same form as those corresponding to an uncoated wire [23, 24], except for additive corrections proportional to σ in the numerator and denominator.

5.3 Scattered and absorbed power

The amplitudes given by equations (5.7)–(5.10) allow us to obtain the electromagnetic field everywhere in space, as well as other quantities of interest, such as the differential, absorption and extinction cross-sections. The derivation of these quantities is identical to that for graphene-free cylinders [22–24].

The interaction of the incident radiation with the graphene-coated wire decreases the power flux of the incident wave and results in scattering and absorption. To calculate the fields in the far region $k_2\rho \to \infty$, the Hankel functions $H_n^{(1)}(k_2\rho)$, $-\infty < n < \infty$ in expansions (5.2) and (5.4) are replaced by their large-argument asymptotic forms [25]

$$H_n^{(1)}(k_2\rho) \to \sqrt{\frac{2}{\pi k_2\rho}} \exp i(k_2\rho - n\pi/2 - \pi/4).$$

Therefore, the scattered field in region 2 can be written as an outgoing cylindrical wave whose amplitude depends on the observation angle ϕ and, using the Poynting vector, it is easy to show that the angular distribution of scattered power per unit solid angle becomes proportional to the modulus squared of the amplitude of that outgoing cylindrical wave. The integral of this angular distribution (the total scattered power), normalized to the incident power, is known as the scattering cross-section [22–24].

While the angular distribution and total scattered power provide information about the far field behavior, the absorbed power provides information about the near fields. As in equation (4.14), the total absorbed power has two contributions: a surface contribution that comes from the dissipation of electric currents in the graphene coating and a volume contribution from the loss in the substrate. The total absorbed power normalized to the incident power is known as the absorption cross-section. The p-polarized total scattering and extinction cross-sections can be calculated using the following expressions

$$Q_{\text{scatt}} = \frac{\pi^2}{k_2} \sum_{n=0}^{\infty} \delta_n |a_n|^2 \tag{5.11}$$

$$Q_{\text{abs}} = \frac{2\pi}{k_2} \operatorname{Im} \sum_{n=0}^{\infty} \delta_n a_n \cos m\phi, \tag{5.12}$$

where δ_m is 1 if $m = 0$, and 2 otherwise. Very similar expressions hold for the cross-sections in the other polarization case.

The p-polarized scattering and absorption cross-sections (per unit length) for a wire with a radius $R = 0.5$ μm, made from a nonplasmonic transparent material ($\varepsilon_1 = 3.9$, $\mu_1 = 1$) in a vaccum ($\mu_2 = \varepsilon_2 = 1$) are shown in figure 5.2 for excitation frequencies in the range between 5 THz (incident wavelength 60 μm) and 21 THz (incident wavelength 14 μm). The Kubo parameters are $T = 300$ K and $\gamma_c = 0.1$ meV, while the number near each curve gives the value of the chemical potential μ_c in eV. The scattering cross-section of the graphene-free cylinder (or of the graphene-coated cylinder for $\mu_c = 0$, gray curve) is also given as a reference and in this case the absorption cross-section vanishes identically. For each value of μ_c a great enhancement in both the scattering and the absorption cross-section spectra can be observed at a well-defined frequency in the lower range, where the scattering cross-section enhancement factor is around four orders of magnitude. A second set of local maxima with lower peaks than the maxima in the first set can be observed in the middle-frequency range (particularly noticeable in the absorption cross-section spectra), whereas a third set of local maxima with even weaker peaks occurs in the higher frequency range, but can only be observed in the absorption cross-section spectra.

The fact that the scattering cross-section spectrum of the graphene-free wire does not show any special feature in the spectral range considered in figure 5.2 indicates that scattering and absorption enhancement is introduced by the graphene coating. As in the case of the peaks of enhanced absorption and local minima in the transmissivity observed in the p polarization case for the strip gratings considered in sections 4.3 and 4.4, the peaks in figure 5.2 are intimately linked with the excitation of localized surface plasmons in the circular wire. As with metallic plasmonic particles, scattering and absorption enhancements occur at a set of discrete frequencies that depend on the size of the wire. However, a very attractive feature of the graphene results is the fact that these frequencies can be tuned by controlling the chemical potential with a gate-voltage. Figure 5.3 illustrates the size dispersion of the first enhancement peaks for graphene-coated dielectric wires, where the core has the same constitutive parameters as those considered in figure 5.2 and $\mu_c = 0.9$ eV for the graphene coating. We observe that the enhancements in the extinction cross-section follow well-defined lines in the $R-\nu$ plane, and each line corresponds to one of the sets of local maxima observed for the low-, middle- and high-frequency ranges considered in figure 5.2.

5.4 Homogeneous problem

While the coefficients a_n and c_n of the field expansions equations (5.3) and (5.4) determine the electromagnetic response of the graphene-coated cylinder when it is illuminated by an incident wave, the complex poles of these coefficients are associated with *localized* surface plasmons, a type of electromagnetic standing wave on the graphene coating that exists without external excitation. In other words,

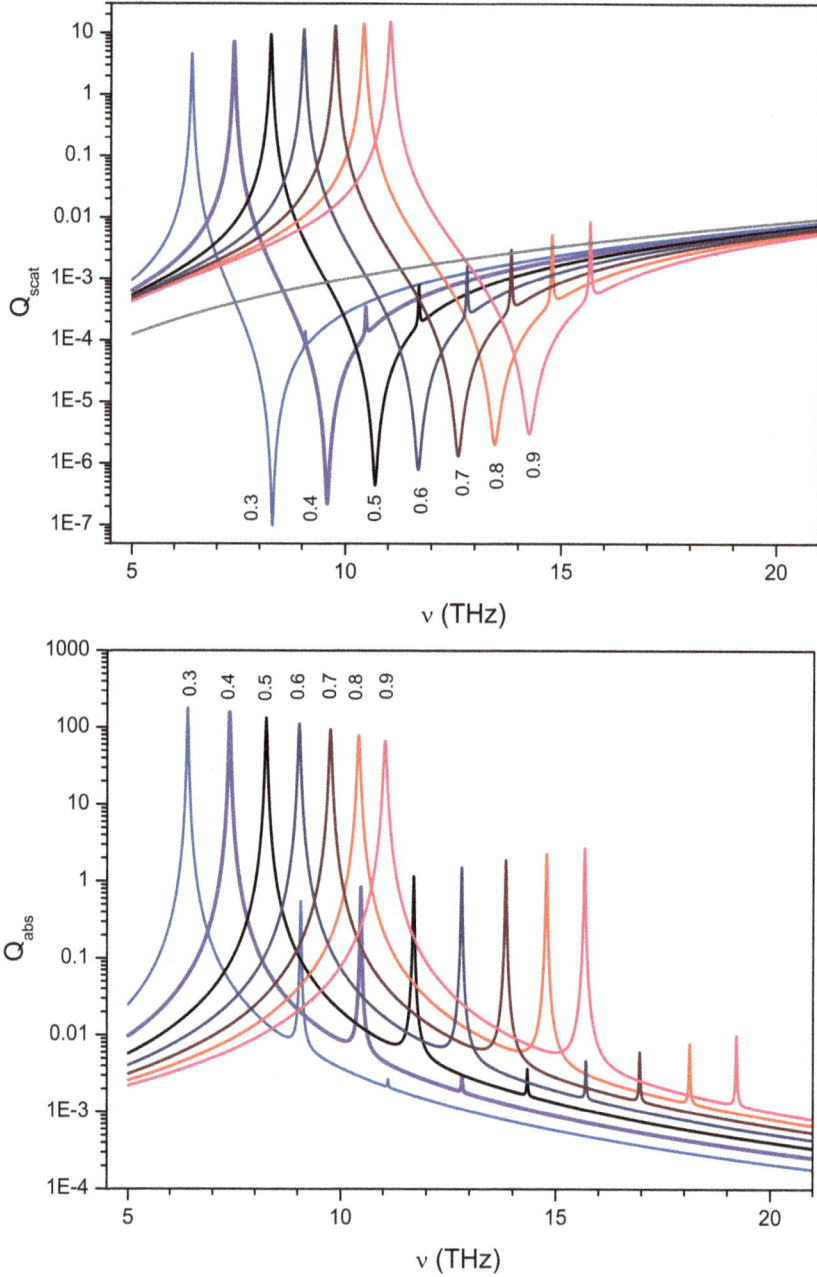

Figure 5.2. Scattering and absorption cross-section spectra for a graphene-coated cylinder, $R = 0.5$ μm, $\varepsilon_1 = 3.9$ and $\mu_1 = \mu_2 = \varepsilon_2 = 1$, illuminated by a p-polarized plane wave. The number near each curve gives the value of the chemical potential μ_c in eV. The gray curve corresponds to the scattering cross-section of the graphene-free cylinder.

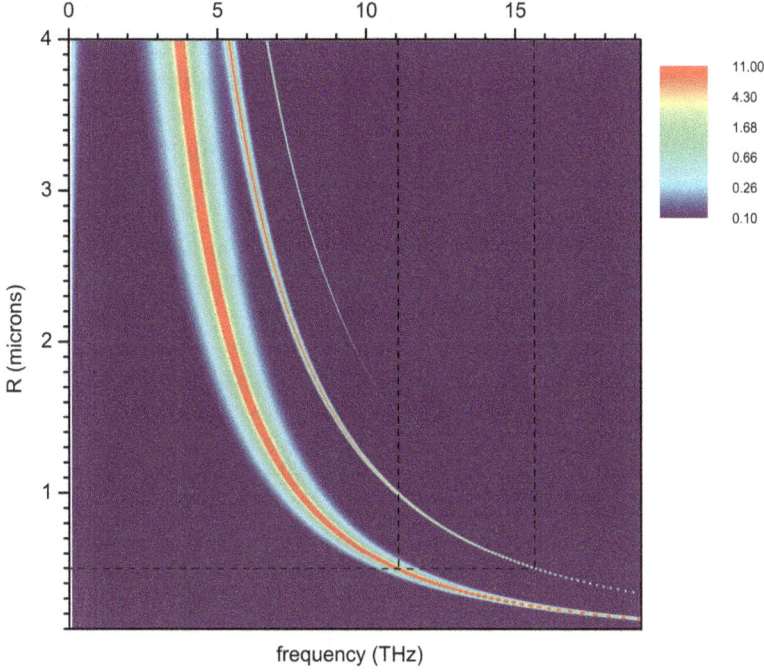

Figure 5.3. Color map for the extinction cross-section of graphene-coated dielectric wires in vacuum ($\mu_2 = \varepsilon_2 = 1$). Vertical axis: wire radius R in microns. Horizontal axis: excitation frequency ν in THz. p polarization, $\mu_c = 0.9$ eV, $\varepsilon_1 = 3.9$ and $\mu_1 = 1$. The black dashed lines correspond to the curves for $\mu_c = 0.9$ eV in figure 5.2, where the first two enhancement peaks are located at $\nu \approx 11.035$ THz and 15.67 THz.

nontrivial electromagnetic solutions exist without an incident wave ($A_0 = 0$) at those frequencies that make zero the denominator in equations (5.7) and (5.9)

$$\frac{\varepsilon_1}{k_1} J_n(x_1) H_n^{'(1)}(x_2) - \frac{\varepsilon_2}{k_2} J_n'(x_1) H_n^{(1)}(x_2) + 4\pi\alpha\,\tilde{\sigma}\frac{c}{\omega} i J_n'(x_1) H_n^{'(1)}(x_2). \qquad (5.13)$$

In section 2.3.1 we found a similar relationship between the poles of the electromagnetic response and the surface modes that exist without external excitation, where the Fresnel reflection and transmission coefficients for a flat graphene interface determine the electromagnetic response when the interface is illuminated by an incident wave, while the complex poles of these coefficients are associated with the propagating surface plasmons that exist without external excitation. However, while for an infinite plane the poles obtained in section 2.3.1 correspond to *propagating* surface plasmons, which cannot couple with incident radiation because of the 2D continuous translational symmetry, for the closed cylindrical surface of the graphene-coated wire considered here, the poles correspond to *localized* surface plasmons, which can couple with incident radiation when the incident frequency coincides with one of the natural frequencies of the standing wave on the graphene coating.

The well-defined lines formed in the $R-\nu$ plane in figure 5.3 indicate the enhancement of the extinction cross-section via the resonant excitation of these localized surface plasmons and therefore coincide with the lines described by the real part of the complex roots of equation (5.13), which is the dispersion relation for the plasmonic eigenmodes of the graphene-coated cylinder. It gives the complex frequencies ω_n, where the nth terms in the cylindrical multipole expansions equations (5.3) and (5.4) become infinitely large. As excitation frequencies are real quantities, the fields given by equations (5.3) and (5.4) never become infinite. However, when the frequency of the excitation is equal to the real part of one of the natural frequencies ω_n, we expect the nth term in the expansions to predominate over the rest. To evidence this relationship, we calculate the spatial distribution of the near field $|f(\rho, \varphi)|$ obtained for incident frequencies $\nu = 11.04$ THz, 15.67 THz and 19.22 THz, that is, the values corresponding to the three enhancement peaks appearing in the scattering and absorption cross-sections (figure 5.2) for $R = 0.5$ μm and $\mu_c = 0.9$ eV (the first two are also indicated by black dashed lines in figure 5.3). As shown by the color map in figure 5.4(a), the near magnetic field for $\nu \approx 11.04$ THz is clearly of an electric dipole nature, as can be expected from the angular dependence of the terms with $n = 1$ in the field expansions. We observe that, in contrast to the case of metallic wires, where the field cannot penetrate into the interior regions and is limited to a surface layer that is approximately one skin depth thick, in the graphene-coated dielectric wire there are regions where the interior field can be greatly enhanced. Analogously, when the excitation frequency is $\nu = 15.67$ THz (figure 5.4(b)) and 19.22 THz (figure 5.4(c)) the spatial distributions of $|f(\rho, \varphi)|$ are of quadrupolar and hexapolar nature, as can be expected from the angular dependence of the terms with $n = 2$ and $n = 3$ in the field expansions, respectively.

5.5 Connection between localized and propagating graphene surface plasmons

When the size of the wire is small enough for the arguments $x_1 = k_1 R$ and $x_2 = k_2 R$ of the Bessel functions in equation (5.13) to be much less than unity, the small-argument asymptotic expansions for Bessel and Hankel functions [25] allow us to obtain the following approximated dispersion relation for localized surface plasmons [26]:

$$\varepsilon_1 + \varepsilon_2 = -4\pi\alpha\,\tilde{\sigma}\frac{c}{\omega}\frac{i}{R}n. \tag{5.14}$$

In the absence of graphene coating, $\tilde{\sigma} = 0$, this approximated relation reduces to the localized surface plasmon resonance condition for a thin wire [15]. It is clear that in graphene-free wires, the existence of localized surface plasmons requires a metallic medium, i.e. $\varepsilon_1\varepsilon_2 < 0$, while for graphene-coated wires in the spectral region where $\operatorname{Im}\sigma > 0$, localized surface plasmons may exist even for dielectric substrates with $\varepsilon_1\varepsilon_2 > 0$.

Taking into account the fact that in the nonretarded regime the wavenumber κ for surface plasmons *propagating* along perfectly flat infinite graphene sheets can be

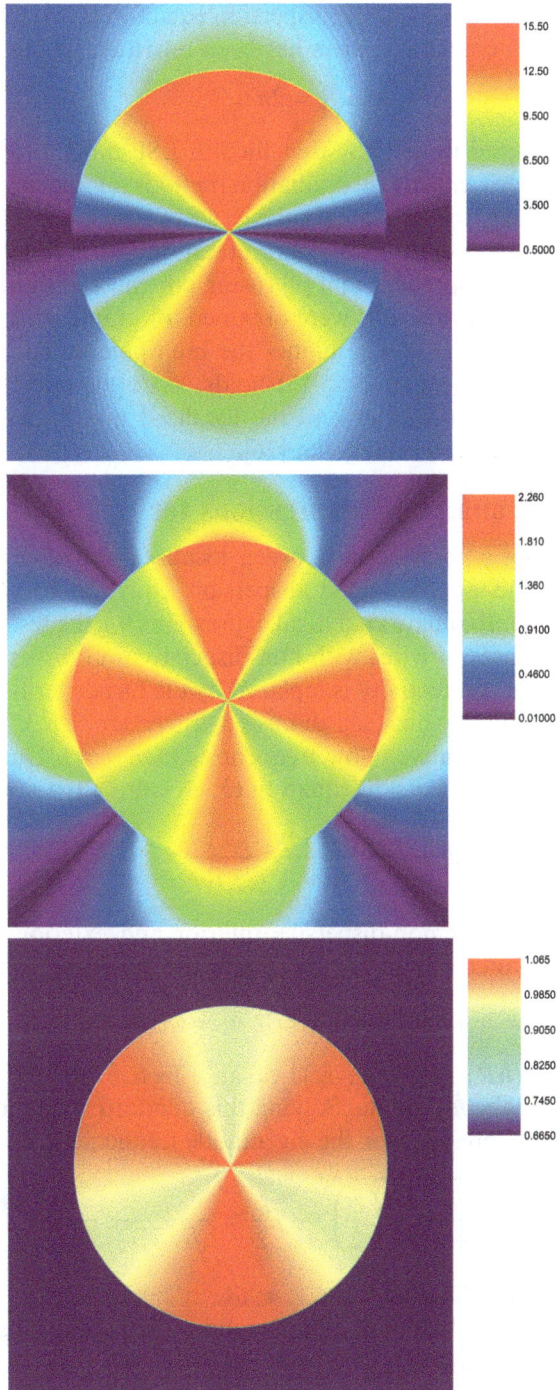

Figure 5.4. Color map for the near field $|f(\rho, \varphi)|$ at frequency values $\nu = 11.04$ THz (a), 15.67 THz (b) and 19.22 THz (c), corresponding to the three enhancement peaks appearing in the scattering and absorption cross-sections for $R = 0.5$ μm and $\mu_c = 0.9$ eV (figure 5.2).

approximated by equation (2.43), the dispersion equation (5.14) for *localized* surface plasmons in graphene-coated cylinders has been rewritten [26] in the form

$$\kappa\, 2\pi R = 2\pi\, n, \tag{5.15}$$

which can be interpreted in the way that the nth LSP mode of a graphene-coated cylinder accommodates exactly n oscillation periods of the propagating surface plasmon corresponding to the flat graphene sheet along the cylinder perimeter. Numerical results that confirm this interpretation are given in [26] (using a separation-of-variables approach) and [9] (using commercial finite element method software). Similar analytical results were obtained for graphene-coated nanospheres in a uniform background [13]. The higher the multipole modal frequency ω_n, the better the approximation equation (5.15) holds, since the effective wavelength of higher multipoles becomes shorter and the localized modes perceive the circular graphene sheet as being increasingly flat [26].

5.6 Approximate analytical expressions for ω_n

Apart from providing a connection between localized and nonlocalized graphene surface plasmons, the approximated dispersion equation (5.14) has also been exploited to obtain approximate analytical expressions for the resonance frequencies ω_n in the particular cases of wires with non-dispersive and dispersive interiors [26]. To do this, the surface conductivity is replaced by the Drude term given by equation (1.37), an approximation which, as discussed in section 1.8.3, is valid for large doping ($\mu_c \gg k_B T$) and relatively low frequencies ($\hbar\omega_n < \mu_c$). Under these assumptions, the approximated dispersion equation (5.14) can be rewritten as

$$\varepsilon_1 + \varepsilon_2 = \frac{4\alpha\mu_c}{\hbar(\omega + i\gamma_c)}\frac{c}{R\omega}\, n. \tag{5.16}$$

For the particular case of non-dispersive interior and exterior media, such as a graphene-coated polymer wire in air, we obtain the quadratic equation

$$(\varepsilon_1 + \varepsilon_2)\omega^2 + i(\varepsilon_1 + \varepsilon_2)\gamma_c\, \omega - \omega_{0n}^2 = 0, \tag{5.17}$$

where $\omega_{0n}^2 = 4\alpha c\mu_c n/(\hbar R)$ plays the role of an effective plasma frequency of the graphene coating for the nth mode. Solving this quadratic equation, we obtain the following analytical expression for the eigenmode frequencies:

$$\omega_n = \sqrt{\frac{\omega_{0n}^2}{\varepsilon_1 + \varepsilon_2} - \left(\frac{\gamma_c}{2}\right)^2} - i\frac{\gamma_c}{2} \approx \frac{\omega_{0n}}{\sqrt{\varepsilon_1 + \varepsilon_2}} - i\frac{\gamma_c}{2}, \tag{5.18}$$

with the second equality holding when $\gamma_c \ll \omega_0$.

For the case of a non-dispersive exterior and a metal-like substrate, we assume that $\varepsilon_1(\omega)$ is represented by

$$\varepsilon_1(\omega) = \varepsilon_\infty - \frac{\omega_p^2}{\omega^2 + i\gamma_m\omega}, \tag{5.19}$$

where ε_∞ is the high-frequency dielectric permittivity, γ_m is the damping constant and ω_p is the plasma frequency. Replacing equation (5.19) into equation (5.16), we obtain

$$\varepsilon_\infty + \varepsilon_2 = \frac{\omega_p^2}{\omega^2 + i\gamma_m\omega} + \frac{\omega_{0n}^2}{\omega^2 + i\gamma_c\omega}, \tag{5.20}$$

which can be rewritten as

$$\omega^2(\varepsilon_\infty + \varepsilon_2) = \frac{\omega_p^2}{1 + ix} + \frac{\omega_{0n}^2}{1 + iy}, \tag{5.21}$$

with $x = \frac{\gamma_m}{\omega}$ and $y = \frac{\gamma_c}{\omega}$. After expanding the right-hand side of equation (5.21) in powers of x and y, and retaining only the first order terms, equation (5.21) can be rewritten as

$$(\varepsilon_\infty + \varepsilon_2)\omega^3 - \left(\omega_p^2 + \omega_{0n}^2\right)\omega + \left(\omega_p^2\gamma_m + \omega_{0n}^2\gamma_c\right)i = 0. \tag{5.22}$$

Separating the frequency into its real and imaginary parts, $\omega = \omega_R - i\omega_I$, $\omega_R > 0$, $\omega_I > 0$, we obtain

$$E\left[\omega_R^3 - 3\omega_R\omega_I^2\right] + iE\left[\omega_I^3 - 3\omega_R^2\omega_I\right] - \Omega^2\omega_R + \Omega^2 i\omega_I + \Gamma i = 0,$$

where we have neglected cubic and quartic terms in ω_I and $E = \varepsilon_\infty + \varepsilon_2$, $\Omega^2 = \omega_p^2 + \omega_{0n}^2$, $\Gamma = \omega_p^2\gamma_m + \omega_{0n}^2\gamma_c$. Finally, the following analytical expressions are obtained for the real and imaginary parts of the eigenmode frequencies

$$\omega_{Rn} = \sqrt{\frac{\omega_p^2 + \omega_{0n}^2}{\varepsilon_\infty + \varepsilon_2}} \qquad \omega_{In} = \frac{\omega_p^2\gamma_m + \omega_{0n}^2\gamma_c}{2\left(\omega_p^2 + \omega_{0n}^2\right)}. \tag{5.23}$$

References

[1] He X, Liu Z, Wang D N, Yang M, Hu T Y and Tian J G 2013 Saturable absorber based on graphene-covered microfiber *IEEE Photon. Technol. Lett.* **25** 1392–4
[2] He X, Zhang X, Zhang H and Xu M 2014 Graphene covered on microfiber exhibiting polarization and polarization-dependent saturable absorption *IEEE J. Sel. Top. Quantum Electron.* **20** 4500107
[3] Wu Y, Yao B, Zhang A, Rao Y, Wang Z, Cheng Y, Gong Y, Zhang W, Chen Y and Chiang K S 2014 Graphene-coated microfiber bragg grating for high-sensitivity gas sensing *Opt. Lett.* **39** 1235–7
[4] Li W *et al* 2014 Ultrafast all-optical graphene modulator *Nano Lett.* **14** 955–9
[5] He X, Xu M, Zhang X and Zhang H 2016 A tutorial introduction to graphene-microfiber waveguide and its applications *Front. Optoelectron.* 1–9
[6] Zhu B, Ren G, Gao Y, Yang Y, Lian Y and Jian S 2014 Graphene-coated tapered nanowire infrared probe: a comparison with metal-coated probes *Opt. Express* **22** 24096
[7] Gao Y, Ren G, Zhu B, Liu H, Lian Y and Jian S 2014 Analytical model for plasmon modes in graphene-coated nanowire *Opt. Express* **22** 24322

[8] Gao Y, Ren G, Zhu B, Wang J and Jian S 2014 Single-mode graphene-coated nanowire plasmonic waveguide *Opt. Lett.* **39** 5909

[9] Huang Z R, Wang L L, Sun B, He M D, Liu J Q, Li H J and Zhai X 2014 A mid-infrared fast-tunable graphene ring resonator based on guided-plasmonic wave resonance on a curved graphene surface *J. Opt.* **16** 105004

[10] Zhao J, Liu X, Qiu W, Ma Y, Huang Y, Wang J-X, Qiang K and Pan J Q 2014 Surface-plasmon-polariton whispering-gallery mode analysis of the graphene monolayer coated InGaAs nanowire cavity *Opt. Express* **22** 5754–61

[11] Xiao T-H, Gan L and Li Z Y 2015 Graphene surface plasmon polaritons transport on curved substrates *Photon. Res.* **3** 300–7

[12] Liu M, Yin X, Ulin-Avila E, Geng B, Zentgraf T, Ju L, Wang F and Zhang X 2011 A graphene-based broadband optical modulator *Nature* **474** 64–7

[13] Christensen T, Jauho A P, Wubs M and Mortensen N 2015 Localized plasmons in graphene-coated nanospheres *Phys. Rev.* B **91** 125414

[14] Maier S A 2007 *Plasmonics: Fundamentals and Applications* (Berlin: Springer)

[15] Novotny L and Hecht B 2012 *Principles of Nano-Optics* 2nd edn (Cambridge: Cambridge University Press)

[16] Hutter E and Fendler J H 2008 Exploitation of localized surface plasmon resonance *Adv. Mater.* **16** 1685–706

[17] Fan X, Zheng W and Singh D J 2014 Light scattering and surface plasmons on small spherical particles *Light: Sci. Appl.* **3** e179

[18] Khatua S, Paulo P P R, Yuan H, Gupta A, Zijlstra P and Orrit M 2014 Resonant plasmonic enhancement of single-molecule fluorescence by individual gold nanorods *ACS Nano* **8** 4440–9

[19] Talley C E, Jackson J B, Oubre C, Grady N K, Hollars C W, Lane S M, Huser T R, Nordlander P and Halas N J 2005 Surface-enhanced raman scattering from individual Au nanoparticles and nanoparticle dimer substrates *Nano Lett.* **5** 1569–74

[20] Li R J, Lin X, Lin S S, Liu X and Chen H S 2015 Tunable deep-subwavelength superscattering using graphene monolayers *Opt. Lett.* **40** 26–33

[21] Riso M, Cuevas M and Depine R A 2015 Tunable plasmonic enhancement of light scattering and absorption in graphene-coated subwavelength wires *J. Opt.* **17** 075001

[22] Velichko E A 2016 Evaluation of a graphene-covered dielectric microtube as a refractive-index sensor in the terahertz range *J. Opt.* **18** 035008

[23] van de Hulst H C 1957 *Light Scattering by Small Particles* (New York: Wiley)

[24] Bohren C F and Huffman D R 1983 *Absorption and Scattering of Light by Small Particles* (New York: Wiley)

[25] Abramowitz A and Stegun I A (ed) 1972 *Handbook of Mathematical Functions with Formulas, Graphs, and Mathematical Tables* 10th edn (New York: Dover)

[26] Riso M, Cuevas M and Depine R A 2016 Complex frequencies and field distributions of localized surface plasmon modes in graphene-coated subwavelength wires *J. Quant. Spectrosc. Radiat.* **114** 26–33

Graphene Optics: Electromagnetic solution of canonical problems

Ricardo A Depine

Appendix

Python scripts

A.1 Python script for figure 1.1

```
 1   # Real and imaginary parts of surface conductivity graphene monolayer
 2   # KUBO MODEL - interband + intraband
 3   # See Figure 1.1 and eqs. (1.36) and (1.38) in
 4   #    Graphene Optics: Electromagnetic solution of canonical problems
 5   #    IOP Concise Physics - 2016
 6   #    Author R. Depine <http://users.df.uba.ar/rdep/>
 7
 8   import numpy as np
 9   import math
10   import matplotlib.pyplot as plt
11   import sigma_Constants      ### Physical Constants
12   import sigmaK
13   pi=math.pi
14   hb=sigma_Constants.hb
15   c=sigma_Constants.c
16   akb=sigma_Constants.akb
17
18   ### Input data
19   TKmu=0.0                 ### T in eV over chemical potential in eV
20   hbaramu=0.1              ### chemical potential in eV
21   hbargama=0.000           ### collision frequency in eV
22   nrof=800                 ### number of points in the curve
23
24   ###   Find frequency range to fit the abscissa range of Figure 1.1
25   ###   0-to-6 for the ratio (hbaromega/hbaramu)
26   OMEGA=np.linspace(0,6,nrof)              ### OMEGA grid
27   hbome1=0.01*hbaramu;   hbome2=6*hbaramu;
28   TK=TKmu*hbaramu/akb                       #Temperature in Kelvin
29   hbaromega=np.linspace(hbome1,hbome2,nrof)     #habaromega grid
30   nuhz=hbaromega/(2*pi*hb)                       #frequency grid Hz
31   lmbd=c / nuhz;                                 #lambda grid microns
32
33   sigmatot, inter, intra = sigmaK.sigma(hbaromega,hbaramu,hbargama,TKmu,nrof)
34
35   ### Save results to file with headers and table
36   f=open('ch1fig1.ext','w')
```

```
37    print >>f, "=================================================================="
38    print >>f, "File ch1fig1.py, script for Figure 1.1 in"
39    print >>f, "Graphene Optics: Electromagnetic solution of canonical problems"
40    print >>f, "=================================================================="
41    print >>f, "TKmu (T in eV over chemical potential in eV) = %g" %TKmu
42    print >>f, "TK (absolute temperature) = %g  Kelvin"  %TK
43    print >>f, "hbargama (collision frequency) = %g eV"  %hbargama
44    print >>f, "hbaramu (chemical potential) = %g  eV"   %hbaramu
45    print >>f, "OMEGA,lmbd,nuhz,Reintra,Imintra,Reinter,Iminter,ReTot,ImTot"
46    ### Table
47    for IJK in range (nrof):
48        print >>f, ("%g %g %g  %g %g %g %g %g %g" %(OMEGA[IJK],lmbd[IJK],
49                        nuhz[IJK],np.real(intra[IJK]),np.imag(intra[IJK]),
50                        np.real(inter[IJK]), np.imag(inter[IJK]),
51                        np.real(sigmatot[IJK]),np.imag(sigmatot[IJK])))
52    f.close()
53
54    ### Show results
55    fig = plt.figure(figsize=(6, 8))
56    plt.plot(OMEGA,np.imag(sigmatot), c='k', ls='-', lw=1., marker='s',
57                        markevery=40 , ms=5)
58    plt.plot(OMEGA[5:nrof],np.real(sigmatot[5:nrof]), c='darkgreen',
59                        ls='-', lw=3., marker='o', markevery=20 )
60    plt.xlabel(r'$\Omega$', fontsize=18)
61    plt.ylabel(r'$\sigma$', fontsize=26)
62    #  plt.title(r'$T/\mu=$%g' %TKmu)
63    if TKmu==0.:
64        plt.axis([0., 6., -1.5, 2.])
65        plt.plot(OMEGA,np.imag(intra), c='b', ls='--', lw=2.)
66        plt.plot(OMEGA,np.imag(inter), c='r', ls='-', lw=2.)
67        plt.legend(['total imaginary','inter & total real','intra imaginary',
68                        'inter imaginary', ], loc='upper right', frameon=False)
69        plt.draw()
70        plt.savefig('Ch01Fig1a.eps', format='eps', dpi=300)
71        plt.savefig('Ch01Fig1a.jpg', format='jpg')
72    else:
73        plt.axis([0., 6., -0.5, 3.])
74        plt.legend(['total imaginary', 'total real'],
75                        loc='upper right', frameon=False)
76        plt.draw()
77        plt.savefig('Ch01Fig1b.eps', format='eps', dpi=300)
78        plt.savefig('Ch01Fig1b.jpg', format='jpg')
```

A.1.1 Physical constants

```
1    ### Physical Constants
2    alfac=1/137.0359            ### fine structure
3    hb=6.58211899*10**(-16)     ### Planck constant hbar in eV*s
4    c = 3*10**(14);             ### light velocity in micron/seg
5    akb=8.6173324E-5;           ### Boltzman constant in eV/K
```

A.1.2 Kubo model

```
1    def sigma (hbw,hbmu,hbgama,TKmu,nrof):
2
3        # Surface conductivity of graphene monolayer
4        # KUBO MODEL - interband + intraband
5        # Eqs. (1.36) and (1.38)  in
6        #   Graphene Optics: Electromagnetic solution of canonical problems
7        #     IOP Concise Physics - 2016
8        # Author R. Depine <http://users.df.uba.ar/rdep/>
9        # hbw     frequency grid in eV
10       # hbmu    chemical potential in eV
11       # hbgama  collision frequency in eV
12       # TKmu    temperature in eV over chemical potential in eV
13       # nrof=len(hbw)
14
15       import sigma_Constants
16       akb=sigma_Constants.akb
17       import math
```

```
18       pi=math.pi
19       import numpy as np
20       inter=np.zeros(nrof)
21
22       if TKmu==0.:
23  ### Intraband conductivity, Fermi-Dirac, T=0K, in units of e**2/hb
24          intra=1j*(1/pi)*abs(hbmu)/(hbw+1j*hbgama);
25  ### Interband conductivity, T=0K, in units of e**2/hb
26          for jj in range (nrof):
27              if(hbw[jj]-2*abs(hbmu)>0):
28                  inter[jj]=0.25
29          inter=inter+ 1j*np.log((hbw-2*abs(hbmu))**2/(hbw+
30                                    2*abs(hbmu))**2)/(4*pi)
31       if TKmu!=0.:
32          TK=TKmu*hbmu/akb
33  ### Intraband conductivity, in units of e**2/hb
34          aux=np.exp(1/(2*TKmu))+np.exp(-1/(2*TKmu))
35          intra=2j*(1/pi)*akb*TK*np.log(aux)/(hbw+1j*hbgama)
36  ### Interband conductivity, in units of e**2/hb
37          inter=0.25*(0.5+np.arctan((hbw-2*hbmu)/(2*TK*akb))/pi-
38                  1j*np.log((hbw+2*hbmu)**2/((hbw-2*hbmu)**2
39                                    +(2*TK*akb)**2))/(2*pi))
40  ### Total (Inter+Intra) conductivity, units of e**2/hb (factor 4)
41  ###
42       sigmatot=inter+intra
43       return sigmatot, inter, intra
```

A.2 Python script for figure 2.2

```
1   # Transmittance vs frequency - free standing graphene
2   # Normal incidence - various temperatures
3   # See Figure 2.2 and eqs. (2.25) and (2.27) in
4   #    Graphene Optics: Electromagnetic solution of canonical problems
5   #    IOP Concise Physics - 2016
6   #    Author R. Depine <http://users.df.uba.ar/rdep/>
7
8   import numpy as np
9   import math
10  import matplotlib.pyplot as plt
11  import sigma_Constants       ### Physical Constants
12  import sigmaK
13  pi=math.pi
14  hb=sigma_Constants.hb
15  c=sigma_Constants.c
16  akb=sigma_Constants.akb
17  alfac=sigma_Constants.alfac
18                          # Input data
19                          # One curve for each Temperature in the list
20  TempK=[5,100,300]       # temperatures in K
21  hbaramu=0.5             # chemical potential in eV
22  hbargama=0.0001         # collision frequency in eV
23  nrof=800                # number of points in each curve
24  hbome1=0.01             # minimum frequency eV
25  hbome2=2                # maximum frequency eV
26
27  hbaromega=np.linspace(hbome1,hbome2,nrof)   #habaromega grid
28  nuhz=hbaromega/(2*pi*hb)     # frequency grid Hz (/10**(12) in THz)
29  lmbd=c / nuhz;               # lambda grid microns
30  OMEGA=hbaromega/hbaramu      # hbaromega/hbaramu
31
32  ### Save results to file with headers and table
33  f=open('ch2fig2.ext','w')
34  print >>f, "=================================================="
35  print >>f, "File ch2fig2.py, script for Figure 2.2 in"
36  print >>f, "Graphene Optics: Electromagnetic solution of canonical problems"
37  print >>f, "=================================================="
38  print >>f, "hbargama (collision frequency) = %g eV"  %hbargama
39  print >>f, "hbaramu (chemical potential) = %g  eV"   %hbaramu
40
41  # Axes and labels for Figure 2.2
42  fig = plt.figure(figsize=(8, 6))
43  plt.axis([0., 2.001, 0.9489, 1.003])
44  plt.xlabel(r'$\hbar\omega$', fontsize=18)
45  plt.ylabel('Transmittance', fontsize=18)
46  plt.text(3.1, 2.45 , r'$\hbar\mu_c$=%g eV'  %hbaramu , fontsize=14)
47  plt.text(3.1, 2.30 , r'$\hbar\gamma$=%g eV'  %hbargama, fontsize=14)
48
```

```
49    IP=0
50
51    for TK in TempK:
52        TKmu=TK*akb/hbaramu      #(T in eV over chemical potential in eV)
53        print >>f, "TKmu (T in eV over chemical potential in eV) = %g" %TKmu
54        print >>f, "TK (absolute temperature) = %g  Kelvin" %TK
55        print >>f, ("TK , hbaromega, OMEGA , lmbd, nuhz, Reintra , Imintra,"
56                           " Reinter, Iminter, ReTot, ImTot , Trans" )
57        sigmatot, inter, intra = sigmaK.sigma(hbaromega,hbaramu,hbargama,TKmu,nrof)
58        CC=1+4*pi*alfac*sigmatot
59        Trans=(np.absolute(2/(CC+1)))**2
60
61        for IJK in range (nrof):
62            print>>f, ("%g %g %g  %g  %g  %g  %g  %g  %g  %g  %g  %g %g" %(TK,
63                            hbaromega[IJK], OMEGA[IJK], lmbd[IJK], nuhz[IJK],
64                            np.real(intra[IJK]),np.imag(intra[IJK]),
65                            np.real(inter[IJK]), np.imag(inter[IJK]),
66                            np.real(sigmatot[IJK]),np.imag(sigmatot[IJK]),
67                            Trans[IJK]))
68
69        if IP == 0:
70            plt.plot(hbaromega,Trans, c='k', ls='-', lw=2, marker='o',
71                                            markevery=99 , ms=5)
72        elif IP == 1:
73            plt.plot(hbaromega,Trans, c='r', ls='-', lw=2., marker='^',
74                                            markevery=80 , ms=6)
75        elif IP == 2:
76            plt.plot(hbaromega,Trans, c='b', ls='-', lw=2., marker='s',
77                                            markevery=69 , ms=4)
78
79        IP=IP+1
80
81    f.close()
82    plt.legend([r'T=5$^\circ$K', r'T=100$^\circ$K', r'T=300$^\circ$K'],
83                                    loc='lower right', frameon=False)
84    #begin annotating arrow
85    plt.annotate(
86        '', xy=(1.75,0.9776), xycoords='data',
87        xytext=(1.75,1.), textcoords='data',
88        arrowprops={'arrowstyle': '<->' })
89    plt.text(1.75, .99, r' $\pi\,\frac{e^2}{\hbar c}$', fontsize=18)
90    #plt.text(1.75, .99, r' $\pi\,e^2 /\hbar c$', fontsize=18)
91    #end annotating arrow
92    plt.draw()
93    plt.savefig('Ch02Fig2.eps', format='eps', dpi=300)
94    plt.savefig('Ch02Fig2.jpg', format='jpg')
```

A.3 Python script for figure 2.4

```
1    # Graphene reflectance, transmittance and absorptance vs angle of incidence
2    # at a fixed frequency. Monolayer between two different transparent media
3    # s and p polarizations. See Figure 2.4 and eqs. (2.15) - (2.18) in
4    #      Graphene Optics: Electromagnetic solution of canonical problems
5    #      IOP Concise Physics - 2016
6    #      Author R. Depine <http://users.df.uba.ar/rdep/>
7    #
8
9    import numpy as np
10   import math
11   import cmath
12   import matplotlib.pyplot as plt
13   pi=math.pi
14   Uno=1+0j
15   import sigma_Constants         ### Physical Constants
16   alfac=sigma_Constants.alfac
17
18                                  ### Input data
19   sigmatot=0.1658+1j*20.3846     ### Kubo nondimensional conductivity
20   eps1=12
21   eps2=4
22   amu1=1.
23   amu2=1.
24   teta0=0.
25   tetaf=89.99
26   ihow=800
27
28   tetagrid=np.linspace(teta0,tetaf,ihow)
29   tetarad=tetagrid*math.pi/180
30
```

```
31    beta1=(eps1*amu1)**(0.5)*np.cos(tetarad)
32    beta2=np.sqrt(Uno*eps2*amu2-eps1*amu1*np.sin(tetarad)**2)
33    rs=(beta1/amu1-beta2/amu2-4*pi*alfac*sigmatot)/(beta1/amu1
34                                   +beta2/amu2+4*pi*alfac*sigmatot)
35    ts=(2*beta1/amu1)/(beta1/amu1+beta2/amu2+4*pi*alfac*sigmatot)
36    rp=(eps2/beta2-eps1/beta1+4*pi*alfac*sigmatot)/(eps2/beta2
37                                   +eps1/beta1+4*pi*alfac*sigmatot)
38    tp=(2*eps2/beta2)/(eps2/beta2+eps1/beta1+4*pi*alfac*sigmatot)
39    aux1=beta2*amu1*np.absolute(ts)**2/(beta1*amu2)
40    aux2=beta2*eps1*np.absolute(tp)**2/(beta1*eps2)
41    RS=np.absolute(rs)**2
42    RP=np.absolute(rp)**2
43    TS=aux1.real
44    TP=aux2.real
45    abS=1.-RS-TS
46    abP=1.-RP-TP
47
48    ### Save results to file with headers and table
49    f=open('ch2fig4.ext','w')
50    print >>f, "=================================================="
51    print >>f, "File ch2fig4.py, script for Figure 2.4 in"
52    print >>f, "Graphene Optics: Electromagnetic solution of canonical problems"
53    print >>f, "=================================================="
54    print >>f, eps1, eps2, amu1, amu2
55    print >>f, "ang, sigmatotR, sigmatotI, TS, RS, abS, TP, RP, abP"
56
57    for IJK in range (ihow):
58        print >>f, "%g %g %g %g %g %g %g %g %g" %(tetagrid[IJK], sigmatot.real,
59                        sigmatot.imag, TS[IJK], RS[IJK], abS[IJK], TP[IJK], RP[IJK],
60                                                                    abP[IJK])
61    f.close()
62
63    ### Show results
64    fig = plt.figure(figsize=(6, 8))
65    plt.xlabel(r'$\theta_0$', fontsize=18)
66    plt.ylabel(r'transmissivity', fontsize=26)
67    plt.plot(tetagrid, TS, c='b', ls='-', lw=2)
68    plt.plot(tetagrid, TP, c='r', ls='--', lw=2)
69    plt.draw()
70    plt.savefig('Ch02Fig4b.eps', format='eps', dpi=300)
71    plt.savefig('Ch02Fig4b.jpg', format='jpg')
72
73    fig = plt.figure(figsize=(6, 8))
74    plt.xlabel(r'$\theta_0$', fontsize=18)
75    plt.ylabel(r'absorption ($10^{-3}$)', fontsize=26)
76    plt.plot(tetagrid, abS/0.001, c='b', ls='-', lw=2)
77    plt.plot(tetagrid, abP/0.001, c='r', ls='--', lw=2)
78    plt.draw()
79    plt.savefig('Ch02Fig4c.eps', format='eps', dpi=300)
80    plt.savefig('Ch02Fig4c.jpg', format='jpg')
81
82    fig = plt.figure(figsize=(6, 8))
83    plt.xlabel(r'$\theta_0$', fontsize=18)
84    plt.ylabel(r'reflectivity', fontsize=26)
85    plt.plot(tetagrid, RS, c='b', ls='-', lw=2)
86    plt.plot(tetagrid, RP, c='r', ls='--', lw=2)
87    plt.draw()
88    plt.savefig('Ch02Fig4a.eps', format='eps', dpi=300)
89    plt.savefig('Ch02Fig4a.jpg', format='jpg')
```

A.4 Python script for figure 2.7

```
1     # Dispersion equation for p-polarized graphene SPPs
2     # Recursvive scheme (2.55) and QE approximation (2.43)
3     # Propagation constant vs frequency
4     # Graphene monolayer between two different transparent media
5     #
6     # See Figure 2.7 and eq. (2.55) in
7     #     Graphene Optics: Electromagnetic solution of canonical problems
8     #     IOP Concise Physics - 2016
9     #     Author R. Depine <http://users.df.uba.ar/rdep/>
10    #
11    import numpy as np
12    import matplotlib.pyplot as plt
13    import math
14    import sigmaK
15    pi=math.pi
16    Uno=1+0j
```

```
17    import sigma_Constants          ### Physical Constants
18    alfac=sigma_Constants.alfac
19    hb=sigma_Constants.hb
20    c=sigma_Constants.c
21    akb=sigma_Constants.akb
22
23                          ### Input data
24
25    TK=300.                 # T, temperature in K
26    hbaramu=0.70            # chemical potential in eV)
27    hbargama=0.000658       # collision frequency in eV (tau=1ps)
28    nrof=200                # number of frequency points
29    nuThz1=1.               # minimum frequency Thz
30    nuThz2=50.              # maximum frequency Thz
31    epsilon2=5.
32    epsilon1=1.
33    deltaepsilon=0.00001
34
35    nuThz=np.linspace(nuThz1,nuThz2,nrof)       #nuhz grid
36    nuhz=nuThz*10**(12)                         #frequency grid THz
37    hbaromega=nuhz*(2*pi*hb)                     #habaromega grid
38    lmbd=c / nuhz;                              #lambda grid microns
39    OMEGA=hbaromega/hbaramu
40    TKmu=TK*akb/hbaramu     #(T in eV over chemical potential in eV)
41    kfree=np.zeros(nrof,dtype=np.complex_)
42    kapp=np.zeros(nrof,dtype=np.complex_)
43    kqe=np.zeros(nrof,dtype=np.complex_)
44
45    sigmatot, inter, intra = sigmaK.sigma(hbaromega,hbaramu,hbargama,TKmu,nrof)
46
47    ### Input data, headers and table to file
48    f=open('ch2fig7.ext','w')
49    print >>f, "================================================================"
50    print >>f, "File ch2fig7.py, script for Figure 2.7 in "
51    print >>f, "Graphene Optics: Electromagnetic solution of canonical problems"
52    print >>f, "================================================================"
53    print >>f, "hbargama (collision frequency) = %g eV"  %hbargama
54    print >>f, "hbaramu (chemical potential) = %g  eV"  %hbaramu
55    print >>f, "TKmu (T in eV over chemical potential in eV) = %g" %TKmu
56    print >>f, "TK (absolute temperature) = %g  Kelvin"  %TK
57    print >>f, "eps1 = %g, eps2= %g " %(epsilon1, epsilon2)
58    print >>f, ("lmbd, nuhz, sigmatotr, sigmatoti, ReKfree, ImKfree,"
59                "ReKcont, ImKcont, ReKqe, ImKqe, control")
60    for IJK in range (nrof):
61        epsilon=epsilon1+deltaepsilon
62        sigadim=sigmatot[IJK]
63        # eq. (2.51) graphene immersed in medium with epsilon1
64        kappa=(epsilon1-1/(2*pi*alfac*sigadim)**2)**0.5
65        kappafree=(epsilon1-1/(2*pi*alfac*sigadim)**2)**0.5
66        kappaqe=1j*(epsilon1+epsilon2)/(4*pi*alfac*sigadim)
67        while epsilon <= epsilon2:
68            R_0_k0=(epsilon1-kappa**2)**0.5
69            if R_0_k0.real+R_0_k0.imag < 0:
70                R_0_k0=-R_0_k0;
71            R_d_k0=(epsilon-kappa**2)**0.5
72            if R_d_k0.real+R_d_k0.imag < 0:
73                R_d_k0=-R_d_k0;
74            A=epsilon*R_0_k0+epsilon1*R_d_k0+4*math.pi*alfac*sigadim*R_d_k0*R_0_k0
75            B=(epsilon/R_0_k0+epsilon1/R_d_k0+4*math.pi*alfac*sigadim
76                                *(R_0_k0/R_d_k0+R_d_k0/R_0_k0))*kappa
77            deltakappa=A/B
78            kappa1=kappa+deltakappa
79            beta2=(epsilon1-kappa1**2)**0.5
80            beta1=(epsilon-kappa1**2)**0.5
81            if beta2.real+beta2.imag < 0:
82                    beta2=-beta2;
83            if beta1.real+beta1.imag < 0:
84                    beta1=-beta1;
85            control=epsilon/beta1+epsilon1/beta2+4*math.pi*alfac*sigadim
86            epsilon=epsilon+deltaepsilon
87            kappa=kappa1                # while closes
88
89        print >>f, (lmbd[IJK], nuhz[IJK], sigmatot.real[IJK], sigmatot.imag[IJK],
90                    kappafree.real, kappafree.imag, kappa.real, kappa.imag,
91                    kappaqe.real, kappaqe.imag, control)
92        kfree[IJK]=kappafree
93        kapp[IJK]=kappa
94        kqe[IJK]=kappaqe
95
96    f.close()
```

A.5 Python script for figures 3.4 and 3.5

```
1    # Attenuated Total Reflection
2    # Graphene layer at the bottom of coupling plate (interface 2-3)
3    # Angular scanning, fixed frequency
4    # TE (s) Reflectivity of a plate (medium 2) that separates a
5    # medium of incidence (1) from a medium of transmission (3)
6    # See Figures 3.4 and 3.5 and eq. (3.10) in
7    #       Graphene Optics: Electromagnetic solution of canonical problems
8    #       IOP Concise Physics - 2016
9    #       Author R. Depine <http://users.df.uba.ar/rdep/>
10   #
11   import numpy
12   import math
13   import cmath
14   import matplotlib.pyplot as plt
15   import sigma_Constants   ### Physical Constants
16   pi=math.pi
17   alfac=sigma_Constants.alfac
18   Uno=1+0j
19
20   # Input data
21   sigadim=0.125-1.j
22   DL = [3.]
23   #DL = [1.3, 2., 3.]        #list with thickness/wavelength values
24   eps1=4.
25   eps2=2.25
26   eps3=2.25
27   amu1=1.
28   amu2=1.
29   amu3=1.
30   nrof=1500                 #(number of points in the curve)
31   teta1=48.; teta2=49.;     #(angular range, degrees)
32
33   tetag=numpy.linspace(teta1,teta2,nrof)          #angular grid degrees
34   tetar=tetag*pi/180                              #angular grid radians
35
36   f=open('ch3fig4.ext','w')
37   fig = plt.figure(figsize=(8, 6))
38   plt.xlabel(r'$\vartheta_0$', fontsize=18)
39   plt.ylabel(r'$R^s$', fontsize=18)
40   plt.axis([teta1, teta2, 0.1, 1.])
41
42   n1c=eps1*amu1
43   n2c=eps2*amu2
44   n3c=eps3*amu3
45   kappa=numpy.sin(tetar)
46   beta1=numpy.sqrt(n1c-n1c*kappa**2)
47   beta2=numpy.sqrt(n2c-Uno*n1c*kappa**2)
48   beta3=numpy.sqrt(n3c-Uno*n1c*kappa**2)
49   r12=(beta1/amu1 - beta2/amu2)/(beta1/amu1 + beta2/amu2)
50   r23=(beta2/amu2 - beta3/amu3 - 4*pi*alfac*sigadim)/(beta2/amu2
51                             + beta3/amu3 + 4*pi*alfac*sigadim)
52
53   for dlam in DL:
54     EX=numpy.exp(2j*2*pi*beta2*dlam)
55     rs=(r12+r23*EX)/(1+r12*r23*EX)
56     RS=numpy.absolute(rs**2)
57     print >>f, ("sigma = (%g , %g) e**2/hbar units"
58                                 %(sigadim.real,sigadim.imag))
59     print >>f, "d/lambda (film width in wavelengths) = %g " %dlam
60     print >>f, "eps1 = %g, amu1= %g " %(eps1, amu1)
61     print >>f, "eps2 = %g, amu2= %g " %(eps2, amu2)
62     print >>f, "eps3r = %g, eps3i = %g, amu3= %g " %(eps3.real, eps3.imag, amu3)
63
64     print >>f, "thetag, RS"
65
66     for IJK in range (nrof):
67           print >>f, "%e %e " %(tetag[IJK], RS[IJK])
68
69     plt.plot(tetag,RS, c='k', ls='-', lw=2)
70     plt.draw()
71
72   plt.savefig('ch3fig4.eps', format='eps', dpi=300)
73   plt.savefig('ch3fig4.jpg', format='jpg')
74   f.close()
```

A.6 Python script for figure 3.6

```
1    # Attenuated Total Reflection
2    # Graphene layer at the bottom of coupling plate (interface 2-3)
3    # Angular scanning, fixed frequency
4    # TM (p) Reflectivity of a plate (medium 2) that separates a
5    # medium of incidence (1) from a medium of transmission (3)
6    # See Figure 3.6 and Section 3.2.2 in
7    #      Graphene Optics: Electromagnetic solution of canonical problems
8    #      IOP Concise Physics - 2016
9    #      Author R. Depine <http://users.df.uba.ar/rdep/>
10   #
11   import numpy
12   import math
13   import cmath
14   import matplotlib.pyplot as plt
15   import sigma_Constants   ### Physical Constants
16   pi=math.pi
17   alfac=sigma_Constants.alfac
18   Uno=1+0j
19
20   # Input data
21   sigadim=0.0486529+10.8258j
22   DL = [0.05, 0.075, 0.1, 0.15]   #list with thickness/wavelength values
23   eps1=16.
24   eps2=2.
25   eps3=1.
26   amu1=1.
27   amu2=1.
28   amu3=1.
29   nrof=1500                      #(number of points in the curve)
30   teta1=20.; teta2=89.99;        #(angular range, degrees)
31
32   tetag=numpy.linspace(teta1,teta2,nrof)          #angular grid degrees
33   tetar=tetag*pi/180                              #angular grid radians
34
35   f=open('ch3fig6.ext','w')
36   fig = plt.figure(figsize=(8, 6))
37   plt.xlabel(r'$\vartheta_0$', fontsize=18)
38   plt.ylabel(r'$R^p$', fontsize=18)
39   plt.axis([teta1, 90., 0., 1.])
40
41   n1c=eps1*amu1
42   n2c=eps2*amu2
43   n3c=eps3*amu3
44   kappa=numpy.sin(tetar)
45   beta1=numpy.sqrt(n1c-n1c*kappa**2)
46   beta2=numpy.sqrt(n2c-Uno*n1c*kappa**2)
47   beta3=numpy.sqrt(n3c-Uno*n1c*kappa**2)
48   r12=(eps2/beta2 - eps1/beta1)/(eps2/beta2 + eps1/beta1)
49   r23=(eps3/beta3 - eps2/beta2 + 4*pi*alfac*sigadim)/(eps3/beta3
50                              + eps2/beta2 + 4*pi*alfac*sigadim)
51
52   for dlam in DL:
53       EX=numpy.exp(2j*2*pi*beta2*dlam)
54       rp=(r12+r23*EX)/(1+r12*r23*EX)
55       RP=numpy.absolute(rp**2)
56       print >>f, ("sigma = (%g , %g) e**2/hbar units"
57                            %(sigadim.real,sigadim.imag))
58       print >>f, "d/lambda (film width in wavelengths) = %g " %dlam
59       print >>f, "eps1 = %g, amu1= %g " %(eps1, amu1)
60       print >>f, "eps2 = %g, amu2= %g " %(eps2, amu2)
61       print >>f, "eps3r = %g, eps3i = %g, amu3= %g " %(eps3.real, eps3.imag, amu3)
62
63       print >>f, "thetag, RP"
64
65       for IJK in range (nrof):
66            print >>f, "%e %e " %(tetag[IJK], RP[IJK])
67
68       plt.plot(tetag, RP, c='k', ls='-', lw=2)
69       plt.draw()
70
71   plt.savefig('ch3fig6.eps', format='eps', dpi=300)
72   plt.savefig('ch3fig6.jpg', format='jpg')
73   f.close()
```

A.7 Python script for figures 3.9 and 3.10

```
1    # Graphene multilayers with N unit cells of the type (AGBG)
2    # Transmissivity and semi--trace of matrix \mbox{\bf P}
3    # TE, spol, fixed angle of incidence, frequency scanning
4    # See Figure 3.9 and 3.10 and eqs. (3.36) and (3.46) in
5    #     Graphene Optics: Electromagnetic solution of canonical problems
6    #     IOP Concise Physics - 2016
7    #     Author R. Depine <http://users.df.uba.ar/rdep/>
8
9    import numpy as np
10   import math
11   import cmath
12   import matplotlib.pyplot as plt
13   import sigma_Constants        ### Physical Constants
14   import sigmaK
15   pi=math.pi
16   hb=sigma_Constants.hb
17   c=sigma_Constants.c
18   akb=sigma_Constants.akb
19   alfac=sigma_Constants.alfac
20
21                     # Input data Graphene
22   TK=300            # T, the range in K was TempK
23   hbaramu=0.25      # chemical potential in eV
24   hbargama=0.0001   # hbargama=0.000658, tau=1ps
25   nrof=20000        # number of frequency points in the curve
26   nuThz1=.25        # minimum frequency Thz
27   nuThz2=15.        # maximum frequency Thz
28
29                     # Input data media
30   G=1.              # G=0, without graphene; G=1, with graphene
31   epsa=2.76
32   amua=1.
33   epsb=7.+1j*0.00
34   amub=1.
35   teta0=0.          # angle of incidence
36   da=11.25          # film width, type a dielectric, microns
37   db=3.75           # film width, type b dielectric, microns
38   N=10              # (AGBG)**N, number of unit cells
39
40   ceror=0.
41   ceroi=0.
42   cero=ceror+1j*ceroi
43   Uno=1.+0j
44
45   Mat=np.matrix([[cero,cero],[cero,cero]])
46   TT=np.matrix([[cero,cero],[cero,cero]])
47   MA1=np.matrix([[cero,cero],[cero,cero]])
48   MA2=np.matrix([[cero,cero],[cero,cero]])
49   MB1=np.matrix([[cero,cero],[cero,cero]])
50   MB2=np.matrix([[cero,cero],[cero,cero]])
51   MP1=np.matrix([[cero,cero],[cero,cero]])
52   MPN=np.matrix([[cero,cero],[cero,cero]])
53
54   lambdanum= np.zeros(nrof,dtype=complex)
55   lambda1= np.zeros(nrof,dtype=complex)
56   lambda2= np.zeros(nrof,dtype=complex)
57   lambdatot= np.zeros(nrof,dtype=complex)
58
59   refle= np.zeros(nrof)
60   trans= np.zeros(nrof)
61   absor= np.zeros(nrof)
62
63   nuThz=np.linspace(nuThz1,nuThz2,nrof)    #nuhz grid
64   nuhz=nuThz*10**(12)                       #frequency grid Hz (/10**(12) in THz)
65   hbaromega=nuhz*(2*pi*hb)                  #habaromega grid
66   lmbd=c / nuhz;                            #lambda grid microns
67   OMEGA=hbaromega/hbaramu
68   TKmu=TK*akb/hbaramu
69
70   sigmatot, inter, intra = sigmaK.sigma(hbaromega,hbaramu,hbargama,TKmu,nrof)
71
72   d=da+db
73   TITAR=teta0*pi/180
74   RIa=math.sqrt(epsa*amua)
75   RIb= cmath.sqrt(epsb*amub)
76
77   ### Save results to file with headers and table
78   f=open('ch3fig9.ext','w')
79   print >>f, "============================================================"
80   print >>f, "File ch3fig9.py, script for Figures 3.9 and 3.10 in "
81   print >>f, "Graphene Optics: Electromagnetic solution of canonical problems"
82   print >>f, "============================================================"
```

```
83     print >>f, "hbargama (collision frequency) = %g eV"  %hbargama
84     print >>f, "hbaramu (chemical potential) = %g  eV"    %hbaramu
85     print >>f, "TKmu (T in eV over chemical potential in eV) = %g" %TKmu
86     print >>f, "TK (absolute temperature) = %g  Kelvin"  %TK
87     print >>f, "angle of incidence = %g (degrees)"   %TITAR
88     print >>f, "eps1 = %g, amu1= %g " %(epsa, amua)
89     print >>f, "eps2r = %g, eps2i = %g, amu2= %g " %(epsb.real, epsb.imag, amub)
90     print >>f, "da = %g (microns), db= %g (microns)" %(da, db)
91     print >>f, "nuThz1 = %g, nuThz2= %g (frequency range, THz)" %(nuThz1, nuThz2)
92     print >>f, "N = %g (number of unit cells)"   %N
93
94     ### headers
95     print >>f, (" hbaromega, lmbd, nuhz, ReTot, ImTot, Trans0, Ref0,"
96                 " Abs0, LnumR, LnumI, L1R, L1I, L2R, L2I " )
97     for IJK in range (nrof):
98         WN0=2*pi/lmbd[IJK]
99         ALFA0=WN0*RIa*math.sin(TITAR)
100        BETAa=WN0*RIa*math.cos(TITAR)
101        AUX2=(WN0*RIb*Uno)**2-ALFA0**2
102        BETAb=cmath.sqrt(AUX2)
103        MA1[0,0]=1.+0.j                                #    MA1
104        MA1[0,1]=1.+0.j
105        MA1[1,0]=-BETAa/amua
106        MA1[1,1]=BETAa/amua
107        MB1[0,0]=1.+0.j                                #    MB1
108        MB1[0,1]=1.+0.j
109        MB1[1,0]=-(BETAb/amub-4*pi*alfac*WN0*sigmatot[IJK])
110        MB1[1,1]= (BETAb/amub+4*pi*alfac*WN0*sigmatot[IJK])
111        MB2[0,0]=cmath.exp(-1j*BETAb*db)              #    MB2
112        MB2[0,1]=cmath.exp(1j*BETAb*db)
113        MB2[1,0]=-BETAb/amub*cmath.exp(-1j*BETAb*db)
114        MB2[1,1]= BETAb/amub*cmath.exp( 1j*BETAb*db)
115        MA2[0,0]=cmath.exp(-1j*BETAa*db)              #    MA2
116        MA2[0,1]=cmath.exp(1j*BETAa*db)
117        MA2[1,0]=-(BETAa/amua-4*pi*alfac*WN0*sigmatot[IJK])*cmath.exp(-1j*BETAa*db)
118        MA2[1,1]= (BETAa/amua+4*pi*alfac*WN0*sigmatot[IJK])*cmath.exp( 1j*BETAa*db)
119        MP1[0,0]=cmath.exp(-1j*BETAa*d)               #    MP1
120        MP1[0,1]=cero
121        MP1[1,0]=cero
122        MP1[1,1]=cmath.exp(1j*BETAa*d)
123        MPN[0,0]=cmath.exp(1j*BETAa*N*d)              #    MPN
124        MPN[0,1]=cero
125        MPN[1,0]=cero
126        MPN[1,1]=cmath.exp(-1j*BETAa*N*d)
127
128    #
129        MAT=MP1*(MA2**(-1)*MB2*MB1**(-1)*MA1)
130        TT=MPN*MAT**(N)
131        refle[IJK]=abs(TT[0,1]/TT[0,0])**2
132        trans[IJK]=1./abs(TT[0,0])**2
133        absor[IJK]=1.-refle[IJK]-trans[IJK]
134
135    #   semi--trace of matrix \mbox{\bf P}
136        lambdanum[IJK]=0.5*(MAT[0,0]+MAT[1,1])
137
138    #   semi--trace of matrix \mbox{\bf P}
139        ca=cmath.cos(BETAa*da)
140        cb=cmath.cos(BETAb*db)
141        sa=cmath.sin(BETAa*da)
142        sb=cmath.sin(BETAb*db)
143        lambda1[IJK]=ca*cb-0.5*((BETAa*amub)/(BETAb*amua)
144                      +(BETAb*amua)/(BETAa*amub))*sa*sb
145        aux1=4*pi*alfac*sigmatot[IJK]*WN0
146        aux2=0.5*aux1**2
147        lambda2[IJK]=(-1j*aux1*(sa*cb*amua/BETAa + sb*ca*amub/BETAb)
148                      -aux2*sa*sb*(amua*amub)/(BETAa*BETAb))
149        lambdatot[IJK]=lambda1[IJK]+lambda2[IJK]
150    #
151        print >>f, "%g %g %g %g %g %g %g %g %g %g %g %g %g %g  %g" %(hbaromega[IJK],
152                    lmbd[IJK], nuhz[IJK], np.real(sigmatot[IJK]),
153                    np.imag(sigmatot[IJK]), refle[IJK], trans[IJK], absor[IJK],
154                    np.real(lambdanum[IJK]), np.imag(lambdanum[IJK]),
155                    np.real(lambda1[IJK]), np.imag(lambda1[IJK]),
156                    np.real(lambda2[IJK]), np.imag(lambda2[IJK]))
157    ### Show results
158    fig1 = plt.figure(figsize=(6, 8))
159    plt.plot(nuhz/10**(12), trans, c='b', ls='-', lw=1)
160    #plt.semilogy(nuhz/10**(12), trans, c='b', ls='-', lw=0.5)
161    plt.draw()
162    plt.savefig('ch3fig9t.eps', format='eps', dpi=300)
163    plt.savefig('ch3fig9t.jpg', format='jpg')
164
165    fig2 = plt.figure(figsize=(6, 8))
166    plt.plot(nuhz/10**(12), np.real(lambda1), c='b', ls='-', lw=2)
167    plt.plot(nuhz/10**(12), np.real(lambda2), c='r', ls='-', lw=2)
168    plt.plot(nuhz/10**(12), np.real(lambda1)+np.real(lambda2), c='k', ls='-', lw=2)
169    plt.plot((0.1, 16.), (1, 1), 'g--')
170    plt.plot((0.1, 16.), (-1, -1), 'g--')
171    plt.draw()
172    plt.savefig('ch3fig9half.eps', format='eps', dpi=300)
173    plt.savefig('ch3fig9half.jpg', format='jpg')
174    f.close()
```

A.8 Python script for figures 4.4 and 4.5

```python
1    # Periodic array of graphene strips - s-polarization
2    # Diffracted orders efficiencies and total absorption
3    # See scheme in Figure 4.3 and results in Figures 4.4 and 4.5 in
4    #    Graphene Optics: Electromagnetic solution of canonical problems
5    #    IOP Concise Physics - 2016
6    #    Author R. Depine <http://users.df.uba.ar/rdep/>
7
8    import numpy as np
9    import math
10   import cmath
11   import matplotlib.pyplot as plt
12   import sigma_Constants        ### Physical Constants
13   import sigmaK
14   pi=math.pi
15
16   ceror=0.;  ceroi=0.; cero=ceror+1j*ceroi; Uno=1.+0j
17
18   hb=sigma_Constants.hb
19   c=sigma_Constants.c
20   akb=sigma_Constants.akb
21   alfac=sigma_Constants.alfac
22
23                         # Input data Graphene
24   TK=300                # temperature in K
25   hbaramu=0.39          # chemical potential in eV
26   hbargama=0.000658     # collision frequency in eV, corresponde to 1ps
27                         # Input data plots
28   nrof=1800             # number of frequency points in the curve
29   nuThz1=1.             # minimum frequency Thz
30   nuThz2=10.            # maximum frequency Thz
31
32                         # Input data for Hwang benchmark
33   eps1=1.
34   eps2=1.
35   amu1=1.
36   amu2=1.
37   teta0=0.              # angle of incidence, degrees
38   DG=70.0               # grating period, microns
39   CG=20.0               # strip width, microns
40   NMAX= 31              # Fourier truncation, symmetric around n=0
41
42   CGDG=CG/DG
43   NCERO=(NMAX+1)/2
44   NCM1=NCERO-1
45   NBMAX=2*NMAX-1
46   N2=2*NMAX
47
48   MM = np.zeros(NBMAX,dtype=np.complex_)
49   MAT= np.zeros((NMAX,NMAX),dtype=np.complex_)
50   Tamp = np.zeros(NMAX,dtype=np.complex_)
51   Ramp = np.zeros(NMAX,dtype=np.complex_)
52   VEC = np.zeros(NMAX,dtype=np.complex_)
53   Absg= np.zeros(nrof,dtype=np.complex_)
54   Trans0= np.zeros(nrof)
55   Ref0= np.zeros(nrof)
56   Abs0= np.zeros(nrof)
57   pottot= np.zeros(nrof)
58
59   nver=np.asarray([x-(NCERO-1) for x in list(range(NMAX))])   # -2 ....2
60   nfur=[x-(NMAX-1) for x in list(range(NBMAX))]                # -4 ....4
61
62   for ii in range(NMAX):
63       MM[ii]=cero
64       for jj in range(NMAX):
65           MAT[ii,jj]=cero
66
67   for ii in range(1,NMAX):
68       MM[ii+NMAX-1]=(np.exp(2j*ii*pi*CGDG)-1)/(2j*ii*pi)
69       MM[NMAX-ii-1]=np.conjugate(MM[ii+NMAX-1])
70
71   MM[NMAX-1]=CGDG
72
73   nuThz=np.linspace(nuThz1,nuThz2,nrof)       #nuhz grid
74   nuhz=nuThz*10**(12)                         #frequency grid Hz (/10**(12) in THz)
75   hbaromega=nuhz*(2*pi*hb)                     #habaromega grid
76   lmbd=c / nuhz;                               #lambda grid microns
77   OMEGA=hbaromega/hbaramu
78   TKmu=TK*akb/hbaramu
79
```

```
80    sigmatot, inter, intra = sigmaK.sigma(hbaromega,hbaramu,hbargama,TKmu,nrof)
81
82    TITAR=teta0*pi/180
83    RIP=np.sqrt(eps1*amu1)
84    RIM=np.sqrt(eps2*amu2)
85
86    ### Save results to file with headers and table
87    f=open('ch4fig4.ext','w')
88    print >>f, "==========================================================="
89    print >>f, "File ch4fig4.py, script for Figures 4.4 and 4.5 in "
90    print >>f, "Graphene Optics: Electromagnetic solution of canonical problems"
91    print >>f, "==========================================================="
92    print >>f, "Graphene strip grating, s-pol"
93    print >>f, "TKmu (T in eV over chemical potential in eV) = %g" %TKmu
94    print >>f, "TK (absolute temperature) = %g  Kelvin"  %TK
95    print >>f, "hbargama (collision frequency) = %g eV"   %hbargama
96    print >>f, "hbaramu (chemical potential) = %g  eV"    %hbaramu
97    print >>f, "angle of incidence = %g (degrees)"   %TITAR
98    print >>f, "eps1 = %g, amu1= %g " %(eps1, amu1)
99    print >>f, "eps2r = %g, eps2i = %g, amu2= %g " %(eps2.real, eps2.imag, amu2)
100   print >>f, "strip width = %g, period= %g (microns)" %(CG, DG)
101   print >>f, "NMAX = %g" %NMAX
102
103   ### headers
104   print >>f, ("hbaromega, lmbd, nuhz, ReTot, ImTot, Trans0, Ref0,"
105                                       "Abs0, Pottot,Absg")
106   for IJK in range (nrof):
107       WN0=2*pi/(lmbd[IJK]/DG)
108       ALFA0=WN0*RIP*np.sin(TITAR)
109       BETA0=WN0*RIP*np.cos(TITAR)
110       ALFA=ALFA0+nver*2*pi
111       AUX1=(WN0*RIP*Uno)**2-ALFA**2
112       BETA1=AUX1**0.5
113       AUX2=(WN0*RIM*Uno)**2-ALFA**2
114       BETA2=AUX2**0.5
115       for mi in range(NMAX):
116           VEC[mi]=cero
117           for ni in range(NMAX):
118               MAT[mi,ni]=4*pi*alfac*WN0*sigmatot[IJK]*MM[ni-mi+NMAX-1]
119           MAT[mi,mi]=MAT[mi,mi]+(BETA1[mi]/amu1+BETA2[mi]/amu2)
120       VEC[NCM1]=2*BETA0/amu1
121
122   #   Linear system for transmitted amplitudes
123       Tamp=np.linalg.solve(MAT, VEC)
124   #   reflected amplitudes
125       Ramp=Uno*Tamp
126       Ramp[NCM1]=Ramp[NCM1]-1.
127       h2=BETA2.real[NCM1]
128       h1=BETA1.real[NCM1]
129       Trans0[IJK]=h2*amu1*np.absolute(Tamp[NCM1]**2)/(h1*amu2)
130       Ref0[IJK]=np.absolute(Ramp[NCM1]**2)
131       pottot[IJK]=0.
132       for ni in range(NMAX):
133           h2=BETA2.real[ni]
134           h1=BETA1.real[ni]
135           pottot[IJK]=pottot[IJK] + (
136                   h2*amu1*np.absolute(Tamp[ni]**2)/(BETA0*amu2)
137               +  h1*np.absolute(Ramp[ni]**2)/(BETA0)   )
138       Abs0[IJK]=1-pottot[IJK]
139
140       aux1=cero
141       for ni in range(NMAX):
142           for mi in range(NMAX):
143               aux1=aux1+Tamp[ni]*np.conj(Tamp[mi])*MM[ni-mi+NMAX-1]
144       Absg[IJK]=4*pi*alfac*amu1*DG*sigmatot[IJK]*aux1/(BETA0*c)
145
146       print >>f, "%g %g %g %g %g %g  %g  %g  %g %g" %(
147                   hbaromega[IJK], lmbd[IJK], nuhz[IJK],
148                   np.real(sigmatot[IJK]),np.imag(sigmatot[IJK]),
149                   Trans0[IJK],Ref0[IJK],
150                   Abs0[IJK],pottot[IJK],np.real(Absg[IJK]) )
151
152   ### Show results, efficiency 0th transmitted
153   fig = plt.figure(figsize=(8, 6))
154   plt.axis([1, 10, -0.01, 1.01])
155   plt.xlabel(r'f (THz)', fontsize=18)
156   plt.ylabel(r'$T_0$', fontsize=18)
157   plt.plot(nuhz/10**(12), Trans0, c='k', ls='-', lw=2)
158   plt.draw()
159   plt.savefig('ch4fig4tr.eps', format='eps', dpi=300)
160   plt.savefig('ch4fig4tr.jpg', format='jpg')
161
162   ### Show results, absorbed
163   fig = plt.figure(figsize=(8, 6))
164   plt.xlabel(r'f (THz)', fontsize=18)
165   plt.ylabel(r'Abs', fontsize=18)
166   plt.semilogy(nuhz/10**(12), Abs0, c='y', ls='-', lw=2)
167   plt.draw()
168   plt.savefig('ch4fig4ab.eps', format='eps', dpi=300)
169   plt.savefig('ch4fig4ab.jpg', format='jpg')
170
171   f.close()
```

A.9 Python script for figures 4.6 and 4.8

```
1    # Periodic array of graphene strips - p-polarization
2    # Diffracted orders efficiencies and total absorption
3    # See scheme in Figure 4.3 and results in Figures 4.6 and 4.8 in
4    #    Graphene Optics: Electromagnetic solution of canonical problems
5    #    IOP Concise Physics - 2016
6    #    Author R. Depine <http://users.df.uba.ar/rdep/>
7
8    import numpy as np
9    import math
10   import cmath
11   import matplotlib.pyplot as plt
12   import sigma_Constants       ### Physical Constants
13   import sigmaK
14   pi=math.pi
15
16   ceror=0.;   ceroi=0.;  cero=ceror+1j*ceroi; Uno=1.+0j
17
18   hb=sigma_Constants.hb
19   c=sigma_Constants.c
20   akb=sigma_Constants.akb
21   alfac=sigma_Constants.alfac
22
23                           # Input data Graphene
24   TK=300                  # temperature in K
25   hbaramu=0.39            # chemical potential in eV
26   hbargama=0.000658       # collision frequency in eV, corresponde to 1ps
27                           # Input data plots
28   nrof=1800              # number of frequency points in the curve
29   nuThz1=1.              # minimum frequency Thz
30   nuThz2=10.             # maximum frequency Thz
31
32                           # Input data for Hwang benchmark
33   eps1=1.
34   eps2=1.
35   amu1=1.
36   amu2=1.
37   teta0=30               # angle of incidence, degrees
38   DG=70.0                # grating period, microns
39   CG=20.0                # strip width, microns
40   NMAX= 501              # Fourier truncation, symmetric around n=0
41
42   CGDG=CG/DG
43   NCERO=(NMAX+1)/2
44   NCM1=NCERO-1
45   NBMAX=2*NMAX-1
46   N2=2*NMAX
47
48   MM = np.zeros(NBMAX,dtype=np.complex_)
49   MAT= np.zeros((NMAX,NMAX),dtype=np.complex_)
50   Tamp = np.zeros(NMAX,dtype=np.complex_)
51   Ramp = np.zeros(NMAX,dtype=np.complex_)
52   VEC = np.zeros(NMAX,dtype=np.complex_)
53   Absg=  np.zeros(nrof,dtype=np.complex_)
54   Trans0= np.zeros(nrof)
55   Ref0= np.zeros(nrof)
56   Abs0= np.zeros(nrof)
57   pottot= np.zeros(nrof)
58
59   nver=np.asarray([x-(NCERO-1) for x in list(range(NMAX))])  # -2 ....2
60   nfur=[x-(NMAX-1) for x in list(range(NBMAX))]              # -4 ....4
61
62   for ii in range(NMAX):
63       MM[ii]=cero
64       for jj in range(NMAX):
65           MAT[ii,jj]=cero
66
67   for ii in range(1,NMAX):
68       MM[ii+NMAX-1]=(np.exp(2j*ii*pi*CGDG)-1)/(2j*ii*pi)
69       MM[NMAX-ii-1]=np.conjugate(MM[ii+NMAX-1])
70
71   MM[NMAX-1]=CGDG
72
73   nuThz=np.linspace(nuThz1,nuThz2,nrof)      #nuhz grid
74   nuhz=nuThz*10**(12)                        #frequency grid Hz (/10**(12) in THz)
75   hbaromega=nuhz*(2*pi*hb)                   #habaromega grid
76   lmbd=c / nuhz;                             #lambda grid microns
77   OMEGA=hbaromega/hbaramu
78   TKmu=TK*akb/hbaramu
79
80   sigmatot, inter, intra = sigmaK.sigma(hbaromega,hbaramu,hbargama,TKmu,nrof)
81
```

```
82   TITAR=teta0*pi/180
83   RIP=np.sqrt(eps1*amu1)
84   RIM=np.sqrt(eps2*amu2)
85
86   ### Save results to file with headers and table
87   f=open('ch4fig8.ext','w')
88   print >>f, "============================================================"
89   print >>f, "File ch4fig8.py, script for Figures 4.6 and 4.8 in "
90   print >>f, "Graphene Optics: Electromagnetic solution of canonical problems"
91   print >>f, "============================================================"
92   print >>f, "Graphene strip grating, p-pol"
93   print >>f, "TKmu (T in eV over chemical potential in eV) = %g" %TKmu
94   print >>f, "TK (absolute temperature) = %g  Kelvin"  %TK
95   print >>f, "hbargama (collision frequency) = %g eV"  %hbargama
96   print >>f, "hbaramu (chemical potential) = %g  eV"   %hbaramu
97   print >>f, "angle of incidence = %g (degrees)"   %TITAR
98   print >>f, "eps1 = %g, amu1= %g " %(eps1, amu1)
99   print >>f, "eps2r = %g, eps2i = %g, amu2= %g " %(eps2.real, eps2.imag, amu2)
100  print >>f, "strip width = %g, period= %g (microns)" %(CG, DG)
101  print >>f, "NMAX = %g" %NMAX
102
103  ### headers
104  print >>f, ("hbaromega, lmbd, nuhz, ReTot, ImTot, Trans0, Ref0,"
105                              "Abs0, Pottot,Absg")
106  for IJK in range (nrof):
107      WNO=2*pi/(lmbd[IJK]/DG)
108      ALFA0=WNO*RIP*np.sin(TITAR)
109      BETA0=WNO*RIP*np.cos(TITAR)
110      ALFA=ALFA0+nver*2*pi
111      AUX1=(WNO*RIP*Uno)**2-ALFA**2
112      BETA1=AUX1**0.5
113      AUX2=(WNO*RIM*Uno)**2-ALFA**2
114      BETA2=AUX2**0.5
115      for mi in range(NMAX):
116          VEC[mi]=cero
117          for ni in range(NMAX):
118              MAT[mi,ni]=BETA1[mi]*BETA2[ni]*4*pi*alfac*(
119                      sigmatot[IJK]*MM[ni-mi+NMAX-1]/(eps1*eps2*WNO))
120          MAT[mi,mi]=MAT[mi,mi]+(BETA1[mi]/eps1+BETA2[mi]/eps2)
121      VEC[NCERO-1]=2*BETA0/eps1
122      Tamp=np.linalg.solve(MAT, VEC)
123      Ramp=-BETA2*Tamp*eps1/(BETA1*eps2)
124      Ramp[NCERO-1]=Ramp[NCERO-1]+1
125      h2=BETA2.real[NCERO-1]
126      h1=BETA1.real[NCERO-1]
127      Trans0[IJK]=h2*eps1*np.absolute(Tamp[NCERO-1]**2)/(h1*eps2)
128      Ref0[IJK]=np.absolute(Ramp[NCERO-1]**2)
129      pottot[IJK]=0.
130      for ni in range(NMAX):
131          h2=BETA2.real[ni]
132          h1=BETA1.real[ni]
133          pottot[IJK]=pottot[IJK] + (
134                  h2*eps1*np.absolute(Tamp[ni]**2)/(BETA0*eps2)
135                  + h1*np.absolute(Ramp[ni]**2)/(BETA0)  )
136      Abs0[IJK]=1-pottot[IJK]
137
138      aux1=cero
139      for ni in range(NMAX):
140          for mi in range(NMAX):
141              aux1=aux1 + ( Tamp[ni] * BETA2[ni] * np.conj(Tamp[mi])
142                      * np.conj(BETA2[mi]) * MM[ni-mi+NMAX-1] )
143      Absg[IJK] = ( 4*pi*alfac*eps1*DG*sigmatot[IJK]*aux1
144                      / (BETA0*c*WNO**2*abs(eps2)**2) )
145
146      print >>f, "%g %g %g %g %g %g  %g  %g  %g %g" %(
147                  hbaromega[IJK], lmbd[IJK], nuhz[IJK],
148                  np.real(sigmatot[IJK]),np.imag(sigmatot[IJK]),
149                  Trans0[IJK],Ref0[IJK],
150                  Abs0[IJK],pottot[IJK],np.real(Absg[IJK]) )
151
152  ### Show results, efficiency 0th transmitted
153  fig = plt.figure(figsize=(8, 6))
154  plt.axis([1, 10, -0.01, 1.01])
155  plt.xlabel(r'f (THz)', fontsize=18)
156  plt.ylabel(r'$T_0$', fontsize=18)
157  plt.plot(nuhz/10**(12), Trans0, c='k', ls='-', lw=2)
158  plt.draw()
159  plt.savefig('ch4fig8tr.eps', format='eps', dpi=300)
160  plt.savefig('ch4fig8tr.jpg', format='jpg')
161
162  ### Show results, absorbed
163  fig = plt.figure(figsize=(8, 6))
164  plt.xlabel(r'f (THz)', fontsize=18)
165  plt.ylabel(r'Abs', fontsize=18)
166  plt.semilogy(nuhz/10**(12), Abs0, c='y', ls='-', lw=2)
167  plt.draw()
168  plt.savefig('ch4fig8ab.eps', format='eps', dpi=300)
169  plt.savefig('ch4fig8ab.jpg', format='jpg')
170
171  f.close()
```

A.10 Python script for figure 4.11

```
1   ##################################################################
2   # Diffraction at a periodically modulated graphene layer
3   # conductivity modulation in the form 1+h*cosKx - p-polarization
4   # Diffracted efficiencies and absorption at fixed frequency
5   # for different angles of incidence
6   #
7   # See Section 4.5 and results in Figure 4.11 in
8   #    Graphene Optics: Electromagnetic solution of canonical problems
9   #    IOP Concise Physics - 2016
10  #    Author R. Depine <http://users.df.uba.ar/rdep/>
11  ##################################################################
12
13  import numpy as np
14  import math
15  import matplotlib.pyplot as plt
16  import sigma_Constants
17  pi=math.pi
18  alfac=sigma_Constants.alfac
19  c=sigma_Constants.c
20
21  # average conductivity
22  sigmatot=0.00647601 + 1j*3.08871
23  # fixed frequency (THz)
24  nuThz=19.8826
25  eps1=1.
26  eps2=1.
27  amu1=1.
28  amu2=1.
29  # minimum angle of incidence (deg)
30  teta0=0.
31  # maximum angle of incidence (deg)
32  tetaf=89.99
33  # number of angles of incidence
34  ihow=800
35  # grating period, microns
36  DG=2.3213
37  # modulation amplitude (see eq. (4.32))
38  hg = 0.1
39  #
40  NMAX= 101
41  NCERO=(NMAX+1)/2
42  NBMAX=2*NMAX-1
43  N2=2*NMAX
44
45  MM = np.zeros(NBMAX,dtype=np.complex_)
46  MAT= np.zeros((NMAX,NMAX),dtype=np.complex_)
47  Tamp = np.zeros(NMAX,dtype=np.complex_)
48  Ramp = np.zeros(NMAX,dtype=np.complex_)
49  VEC = np.zeros(NMAX,dtype=np.complex_)
50
51  ceror=0.; ceroi=0.; cero=ceror+1j*ceroi; Uno=1.+0j
52  Trans0array = np.zeros(ihow)
53  tetarray = np.zeros(ihow)
54  Abs0array = np.zeros(ihow)
55
56  nver=np.asarray([x-(NCERO-1) for x in list(range(NMAX))])
57  nfur=[x-(NMAX-1) for x in list(range(NBMAX))]
58
59  # frequency Hz (/10**(12) in THz)
60  nuhz=nuThz*10**(12)
61  # lambda microns
62  lmbd=c / nuhz
63  RIP=np.sqrt(eps1*amu1)
64  RIM=np.sqrt(eps2*amu2)
65  WN0=2*pi/(lmbd/DG)
66
67  ### Save results to file with headers and table
68  f=open('ch4fig11.ext','w')
69  print >>f, "=============================================================="
70  print >>f, "File ch4fig11.py, script for Figure 4.11 in "
71  print >>f, "Graphene Optics: Electromagnetic solution of canonical problems"
72  print >>f, "=============================================================="
73  print >>f, "Graphene grating with sinusoidal modulation, p-pol"
74  print >>f, ("tetag, lmbd, nuhz, np.real(sigmatot), np.imag(sigmatot),"
75                                      "Trans0, Ref0, Abs0, pottot")
76  print >>f, "nuhz = %g, lmbd = %g, DG= %g, hg= %g " %(nuhz, lmbd, DG, hg)
77  print >>f, "eps1 = %g, amu1= %g " %(eps1, amu1)
78  print >>f, "eps2 = %g, amu2= %g " %(eps2, amu2)
79  print >>f, "NMAX = %g" %NMAX
80  print >>f, "Re Sigma*Tot = %g, Im Sigma*Tot = %g (in units of alfa*c)" \
81              %(sigmatot.real, sigmatot.imag)
82
```

```
83    for ii in range(NMAX):
84        MM[ii]=cero
85        for jj in range(NMAX):
86            MAT[ii,jj]=cero
87
88    for ii in range(1,NMAX):
89        MM[ii+NMAX-1]=cero
90        MM[NMAX-ii-1]=np.conjugate(MM[ii+NMAX-1])
91
92    MM[NMAX-1]=1.0
93    MM[NMAX-2]=hg*0.5
94    MM[NMAX]=hg*0.5
95
96    for IJK in range(ihow):
97        tetag=teta0+IJK*(tetaf-teta0)/(ihow-1)
98        TITAR=tetag*pi/180.
99        tetarray[IJK]=tetag
100       ALFA0=WN0*RIP*np.sin(TITAR); BETA0=WN0*RIP*np.cos(TITAR)
101       ALFA=ALFA0+nver*2*pi
102       AUX1=(WN0*RIP*Uno)**2-ALFA**2
103       BETA1=AUX1**0.5
104       AUX2=(WN0*RIM*Uno)**2-ALFA**2
105       BETA2=AUX2**0.5
106
107       for mi in range(NMAX):
108           VEC[mi]=cero
109           for ni in range(NMAX):
110               MAT[mi,ni] = (BETA1[mi]*BETA2[ni]*4*pi*alfac*sigmatot
111                           * MM[ni-mi+NMAX-1]/(eps1*eps2*WN0) )
112           MAT[mi,mi]=MAT[mi,mi]+(BETA1[mi]/eps1+BETA2[mi]/eps2)
113       VEC[NCERO-1]=2*BETA0/eps1
114       Tamp=np.linalg.solve(MAT, VEC)
115       Ramp=-BETA2*Tamp*eps1/(BETA1*eps2)
116       Ramp[NCERO-1]=Ramp[NCERO-1]+1
117
118       h2=BETA2.real[NCERO-1]
119       h1=BETA1.real[NCERO-1]
120       Trans0=h2*eps1*np.absolute(Tamp[NCERO-1]**2)/(h1*eps2)
121       Trans0array[IJK]=Trans0
122       Ref0=np.absolute(Ramp[NCERO-1]**2)
123       pottot=0.
124       for ni in range(NMAX):
125           h2=BETA2.real[ni]
126           h1=BETA1.real[ni]
127           pottot=pottot+(h2*eps1*np.absolute(Tamp[ni]**2)/(BETA0*eps2)
128                         + h1*np.absolute(Ramp[ni]**2)/(BETA0) )
129       Abs0=1-pottot
130       Abs0array[IJK]=Abs0
131       print IJK, tetag,Abs0
132
133       print >>f, "%g %g %g %g %g %g %g %g %g" %(tetag, lmbd, nuhz,
134                   np.real(sigmatot), np.imag(sigmatot),
135                                   Trans0,Ref0,Abs0,pottot )
136
137   ### Show results, efficiency 0th transmitted
138   fig = plt.figure(figsize=(8, 6))
139   plt.xlabel(r'$\theta_0$ (deg)', fontsize=18)
140   plt.ylabel(r'$T_0$', fontsize=18)
141   plt.plot(tetarray, Trans0array, c='k', ls='-', lw=2)
142   plt.draw()
143   plt.savefig('ch4fig11tr.eps', format='eps', dpi=300)
144   plt.savefig('ch4fig11tr.jpg', format='jpg')
145
146   ### Show results, absorbed
147   fig = plt.figure(figsize=(8, 6))
148   plt.xlabel(r'$\theta_0$ (deg)', fontsize=18)
149   plt.ylabel(r'Abs', fontsize=18)
150   plt.semilogy(tetarray, Abs0array, c='y', ls='-', lw=2)
151   plt.draw()
152   plt.savefig('ch4fig11ab.eps', format='eps', dpi=300)
153   plt.savefig('ch4fig11ab.jpg', format='jpg')
154
155   f.close()
```

A.11 Python script for figure 5.2

```
1     # Graphene coated circular cylinder, p-pol
2     # Scattering and absorption cross-sections vs incident frequency
3     # See scheme in Figure 5.1 and results in Figure 5.2 in
4     #     Graphene Optics: Electromagnetic solution of canonical problems
5     #     IOP Concise Physics - 2016
6     #     Author R. Depine <http://users.df.uba.ar/rdep/>
7
8     import numpy as np
9     import scipy.special as sp
10    import math
```

```
11   import cmath
12   import matplotlib.pyplot as plt
13   import sigma_Constants        ### Physical Constants
14   import sigmaK
15   pi=math.pi
16
17   #########################################################
18   #                       Input data
19   #########################################################
20
21   NMAX=15               # from -NMAX to NMAX
22   TK=300                # temperature in K
23   hbaramu=0.9           # chemical potential in eV
24   hbargama=0.0001       # collision frequency in eV
25
26   nrof=4000             # number of frequency points in the curve
27   nuThz1=5.             # minimum frequency Thz
28   nuThz2=20.            # maximum frequency Thz
29
30   eps1=3.9              # interior
31   amu1=1.
32   eps2=1.               # exterior
33   amu2=1.
34   R=0.5                 # wire radius, microns
35
36   #########################################################
37
38   N1=NMAX+1
39   N2=NMAX+2
40   orders=range(N2)
41
42   DCJ1 = np.zeros(N1,dtype=np.complex_)
43   DCJ2 = np.zeros(N1,dtype=np.complex_)
44   DCH2 = np.zeros(N1,dtype=np.complex_)
45   AN = np.zeros(N1,dtype=np.complex_)        # exterior
46   CN = np.zeros(N1,dtype=np.complex_)        # interior
47   Qabs = np.zeros(nrof)
48   Qscat = np.zeros(nrof)
49
50   ceror=0.;   ceroi=0.; cero=ceror+1j*ceroi; Uno=1.+0j
51
52   hb=sigma_Constants.hb
53   c=sigma_Constants.c
54   akb=sigma_Constants.akb
55   alfac=sigma_Constants.alfac
56
57   RI1 = (eps1*amu1*Uno)**0.5                 # interior
58   RI2 = (eps2*amu2*Uno)**0.5                 # exterior
59
60   nuThz=np.linspace(nuThz1,nuThz2,nrof)      # nuhz grid
61   nuhz=nuThz*10**(12)                        # frequency grid Hz (/10**(12) in THz)
62   hbaromega=nuhz*(2*pi*hb)                   # habaromega grid
63   lmbd=c / nuhz;                             # lambda grid microns
64   OMEGA=hbaromega/hbaramu
65   TKmu=TK*akb/hbaramu
66
67   sigmatot, inter, intra = sigmaK.sigma(hbaromega,hbaramu,hbargama,TKmu,nrof)
68
69   ### Save results to file with headers and table
70   f=open('ch5fig2.ext','w')
71   print >>f, "================================================================="
72   print >>f, "File ch5fig2.py, script for Figure 5.2 in "
73   print >>f, "Graphene Optics: Electromagnetic solution of canonical problems"
74   print >>f, "================================================================="
75   print >>f, "Scattering and absorption cross-sections"
76   print >>f, "Graphene coated circular cylinder, p-pol"
77   print >>f, "TKmu (T in eV over chemical potential in eV) = %g" %TKmu
78   print >>f, "TK (absolute temperature) = %g  Kelvin" %TK
79   print >>f, "hbargama (collision frequency) = %g eV" %hbargama
80   print >>f, "hbaramu (chemical potential) = %g  eV"  %hbaramu
81   print >>f, "eps1 = %g, amu1= %g " %(eps1, amu1)
82   print >>f, "eps2r = %g, eps2i = %g, amu2= %g " %(eps2.real, eps2.imag, amu2)
83   print >>f, "cylinder radius R = %g, microns)" %R
84   print >>f, "NMAX = %g" %NMAX
85
86   ### headers
87   print >>f, ("hbaromega, lmbd, nuhz, sigmaR, sigmaI, Qabs, Qscat")
88
89   for IJK in range (nrof):
90       sigma=sigmatot[IJK]
91       WN0=2*pi/(lmbd[IJK])
92       WN1=WN0*RI1
93       WN2=WN0*RI2
94       x1=WN1*R
95       x2=WN2*R
96       CJ1=sp.jn(orders,x1)
97       CJ2=sp.jn(orders,x2)
98       CH2=sp.hankel1(orders,x2)
99       DCJ1[0]=-CJ1[1]
100      DCJ2[0]=-CJ2[1]
101      DCH2[0]=-CH2[1]
102      for n in range(1,NMAX):
```

```
103            DCJ1[n]=.5*(CJ1[n-1]-CJ1[n+1])
104            DCJ2[n]=.5*(CJ2[n-1]-CJ2[n+1])
105            DCH2[n]=.5*(CH2[n-1]-CH2[n+1])
106
107     for n in range(NMAX):
108            den = ( WN2*eps1*CJ1[n]*DCH2[n] - WN1*eps2*DCJ1[n]*CH2[n]
109                  + 4.0j*pi*alfac*sigma*WN1*WN2*DCJ1[n]*DCH2[n]/WN0 )
110            annum = -1j**n * (
111                     WN2*eps1*CJ1[n]*DCJ2[n] - WN1*eps2*DCJ1[n]*CJ2[n]
112                  + 4.0j*pi*alfac*sigma*WN1*WN2*DCJ1[n]*DCJ2[n]/WN0 )
113            cnnum = WN2 * eps1 * 1j**n * (
114                     CJ2[n]*DCH2[n] - DCJ2[n]*CH2[n] )
115            AN[n] = annum/den
116            CN[n] = cnnum/den
117
118     POTINC=amu2*R*WN0/(4.*pi*WN2)
119     auxa=np.absolute(DCJ1[0]*CN[0])**2
120     auxt=np.absolute(AN[0])**2
121
122     for n in range(1,NMAX):
123            auxa=auxa + 2.*np.absolute(DCJ1[n]*CN[n])**2
124            auxt=auxt + 2.*np.absolute(AN[n])**2
125
126     auxa=auxa*(2.*pi*R*alfac*WN0)/(WN1**2*POTINC)
127     Qabs[IJK]=auxa.real
128     auxt=2.*auxt/(WN2*R)**2
129     Qscat[IJK]=auxt.real
130
131     print >>f, "%g %g %g %g %g %g  %g " %(
132                  hbaromega[IJK], lmbd[IJK], nuhz[IJK],
133                  np.real(sigmatot[IJK]),np.imag(sigmatot[IJK]),
134                  Qabs[IJK],Qscat[IJK] )
135
136 ### Show results, efficiency 0th transmitted
137 fig = plt.figure(figsize=(8, 6))
138 plt.xlabel(r'f (THz)', fontsize=18)
139 plt.ylabel(r'$Q_{abs}$', fontsize=18)
140 plt.semilogy(nuhz/10**(12), Qabs, c='k', ls='-', lw=2)
141 plt.draw()
142 plt.savefig('ch5fig2a.eps', format='eps', dpi=300)
143 plt.savefig('ch5fig2a.jpg', format='jpg')
144 ### Show results, absorbed
145 fig = plt.figure(figsize=(8, 6))
146 plt.xlabel(r'f (THz)', fontsize=18)
147 plt.ylabel(r'$Q_{scat}$', fontsize=18)
148 plt.semilogy(nuhz/10**(12), Qscat, c='y', ls='-', lw=2)
149 plt.draw()
150 plt.savefig('ch5fig2b.eps', format='eps', dpi=300)
151 plt.savefig('ch5fig2b.jpg', format='jpg')
152
153 f.close()
```